UNCONVENTIONAL MANUFACTURING PROCESS

UNCONVENTIONAL MANUFACTURING PROCESS

M K Singh
BTech., MSc., PhD
Former Professor and Head
Department of Mechanical Engineering
Hindustan College of Science & Technology
Farah, Mathura
UP

Publishing Globally

NEW AGE INTERNATIONAL (P) LIMITED, PUBLISHERS
LONDON • NEW DELHI • NAIROBI
Bangalore • Chennai • Cochin • Guwahati • Hyderabad • Kolkata • Lucknow • Mumbai
Visit us at www.newagepublishers.com

Published by New Age International (P) Ltd., Publishers
First Edition: 2008
Reprint: 2019

GLOBAL OFFICES

- **New Delhi** **NEW AGE INTERNATIONAL (P) LIMITED, PUBLISHERS**
7/30 A, Daryaganj, New Delhi-110002, (INDIA)
Tel.: (011) 23253771, 23253472, **Telefax:** 23267437, 43551305
E-mail: contactus@newagepublishers.com • Visit us at www.newagepublishers.com
- **London** **NEW AGE INTERNATIONAL (UK) LTD.**
27 Old Gloucester Street, London, WC1N 3AX, UK
E-mail: info@newacademicscience.co.uk • Visit us at www.newacademicscience.co.uk
- **Nairobi** **NEW AGE GOLDEN (EAST AFRICA) LTD.**
Ground Floor, Westlands Arcade, Chiromo Road (Next to Naivas Supermarket)
Westlands, Nairobi, KENYA, **Tel.:** 00-254-713848772, 00-254-725700286
E-mail: kenya@newagepublishers.com

BRANCHES

- **Bangalore** 37/10, 8th Cross (Near Hanuman Temple), Azad Nagar, Chamarajpet, Bangalore- 560 018
Tel.: (080) 26756823, **Telefax:** 26756820, **E-mail: bangalore@newagepublishers.com**
- **Chennai** 26, Damodaran Street, T. Nagar, Chennai-600 017, **Tel.:** (044) 24353401, **Telefax:** 24351463
E-mail: chennai@newagepublishers.com
- **Cochin** CC-39/1016, Carrier Station Road, Ernakulam South, Cochin-682 016
Tel.: (0484) 2377303, **Telefax:** 4051304, **E-mail: cochin@newagepublishers.com**
- **Guwahati** Hemsen Complex, Mohd. Shah Road, Paltan Bazar, Near Starline Hotel, Guwahati-781 008
Tel.: (0361) 2513881, **Telefax:** 2543669
E-mail: guwahati@newagepublishers.com
- **Hyderabad** 105, 1st Floor, Madhiray Kaveri Tower, 3-2-19, Azam Jahi Road, Near Kumar Theater, Nimboliadda Kachiguda, Hyderabad-500 027, **Tel.:** (040) 24652456, **Telefax:** 24652457
E-mail: hyderabad@newagepublishers.com
- **Kolkata** RDB Chambers (Formerly Lotus Cinema) 106A, 1st Floor, S N Banerjee Road, Kolkata-700 014
Tel.: (033) 22273773, **Telefax:** 22275247
E-mail: kolkata@newagepublishers.com
- **Lucknow** 16-A, Jopling Road, Lucknow-226 001, **Tel.:** (0522) 2209578, 4045297, **Telefax:** 2204098
E-mail: lucknow@newagepublishers.com
- **Mumbai** 142C, Victor House, Ground Floor, N.M. Joshi Marg, Lower Parel, Mumbai-400 013
Tel.: (022) 24927869, **Telefax:** 24915415, **E-mail: mumbai@newagepublishers.com**
- **New Delhi** 22, Golden House, Daryaganj, New Delhi-110 002, **Tel.:** (011) 23262368, 23262370
Telefax: 43551305, **E-mail: sales@newagepublishers.com**

ISBN: 978-81-224-2244-3
C-18-12-11692

Printed in India at Saras Graphics, Rai, Haryana.
Typeset at Abro Enterprises, Delhi.

NEW AGE INTERNATIONAL (P) LIMITED, PUBLISHERS
7/30 A, Daryaganj, New Delhi-110002
Visit us at **www.newagepublishers.com**
(CIN: U74899DL1966PTC004618)

Dedication

To the Memory of My Parents

and

To My Elder Brother Sri Awdhesh Singh whose Inspiration and Dedication Brought Me to This Level.

Preface

In present age the development of technology in the field of processes and materials is so rapid that we can consider it as another technological revolutions. In this revolution the discovery and development of new materials and alloys of characteristics suitable for aerospace, nuclear and other such new industries is taking place. The advent of these new materials which are high strength temp. resistant material as well as increasingly complex part configurations has resulted in the creation of a new, unique family of manufacturing processes known as unconventional manufacturing processes.

The unconventional processes differ from conventional processes (e.g., drilling, turning and stamping) either by utilizing energy in novel ways or by applying forms of energy here tofore unused for the purpose of manufacturing. High velocity material jets, pubed magnetic fields, light beams and electrochemical reactions are a few of these new tools currently used to perform operations such as cutting, welding, deburring and forming. There is a wide variety of much unconventional processes, each designed and developed to cope with specific needs of modern industry catering to the demands of aerospace, nuclear power and other sophisticated developments. The basic theories, the requirements of power and knowledge of design, technology and equipment for these processes are varied in character and diverse in nature. Rapid changes in these areas are still taking place but there is no single source where information on these unconventional manufacturing processes may be readily available.

A number of technical papers and reports have addressed various unconventional manufacturing processes. However, it is impossible for those who are interested to be aware of all these developments concerning this very broad and rapidly changing subject. Few books are also written on this subject but these are incomplete in scope.

In this book an attempt has been made to present the historical background and characteristic features of the unconventional manufacturing processes now available such as abrasive jet machining, ultrasonic machining, chemical machining, electro-chemical grinding, electro-chemical machining, electrical discharge machining, electron beam machining, laser beam machining, plasma arc machining. Basic principle of each process, equipments required, process parameters and its possible fields of applications are incorporated. This book does not cover only unconventional machining processes but also includes the unconventional forming and joining processes.

Specifically this book provides engineers, designers, students and academia with a single source of information explaining what these unconventional processes are, what they are capable of doing and how they benefit manufacturers.

—M.K. Singh

Contents

Introduction of Unconventional Manufacturing Process

1.0 INTRODUCTION

Since beginning of the human race, people have evolved tools and energy sources to power these tools to meet the requirements for making the life more easier and enjoyable.

In the early stage of mankind, tools were made of stone for the item being made. When iron tools were invented, desirable metals and more sophisticated articles could be produced. In twentieth century products were made from the most durable and consequently, the most unmachinable materials. In an effort to meet the manufacturing challenges created by these materials, tools have now evolved to include materials such as alloy steel, carbide, diamond and ceramics.

A similar evolution has taken place with the methods used to power our tools. Initially, tools were powered by muscles; either human or animal. However as the powers of water, wind, steam and electricity were harnessed, mankind was able to further extend manufacturing capabilities with new machines, greater accuracy and faster machining rates.

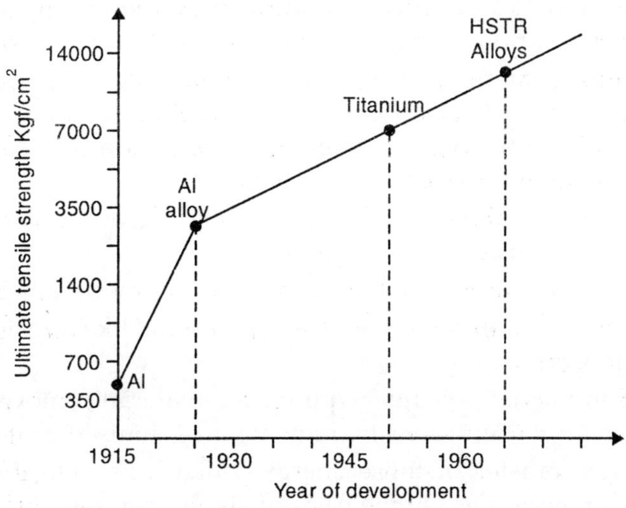

Fig. 1.1. *Trend of increase of material strength*

Every time new tools, tool materials, and power sources are utilized, the efficiency and capabilities of manufacturers are greatly enhanced. Since 1940's, a revolution in manufacturing has been taking place that once again allows manufactuers to meet the demands imposed by increasingly sophisticated designs and durable but in many cases nearly unmachinable, materials.

In the figure 1.1, Merchant had displayed the gradual increase in strength of material with year wise development of material in aerospace industry. This manufacturing revolution is now, as it has been in the past, centered on the use of new tools and new forms of energy. The result has been the introduction of new manufacturing processes used for material removal, forming and joining, known today as non-traditional manufacturing processes.

The conventional manufacturing processes in use today for material removal primarily rely on electric motors and hard tool materials to perform tasks such as sawing, drilling and broaching. Conventional forming operations are performed with the energy from electric motors, hydraulics and gravity. Likewise, material joining is conventionally accomplished with thermal energy sources such as burning gases and electric arcs.

In contrast, non-traditional manufacturing processes harness energy sources considered unconventional by yesterday's standards. Material removal can now be accomplished with electrochemical reaction, high temperature plasmas and high-velocity jets of liquids and abrasives. Materials that in the past have been extremely difficult to form, are now formed with magnetic fields, explosives and the shock waves from powerful electric sparks. Material-joining capabilities have been expanded with the use of high-frequency sound waves and beams of electrons and coherent light.

During the last 55 years, over 20 different non-traditional manufacturing processes have been invented and successfully implemented into production.

1.1 CLASSIFICATION OF UNCONVENTIONAL MANUFACTURING PROCESSES

The non-conventional manufacturing processes are not affected by hardness, toughness or brittleness of material and can produce any intricate shape on any workpiece material by suitable control over the various physical parameters of the processes.

The non-conventional manufacturing processes may be classified on the basis of type of energy namely, mechanical, electrical, chemical, thermal or magnetic, apply to the workpiece directly and have the desired shape transformation or material removal from the work surface by using different scientific mechanism.

Thus, these non-conventional processes can be classified into various groups according to the basic requirements which are as follows :

(*i*) Type of energy required, namely, mechanical, electrical, chemical etc.

(*ii*) Basic mechanism involved in the processes, like erosion, ionic dissolution, vaporisation etc.

(*iii*) Source of immediate energy required for material removal, namely, hydrostatic pressure, high current density, high voltage, ionised material, etc.

(*iv*) Medium for transfer of those energies, like high velocity particles, electrolyte, electron, hot gases, etc. On the basis of above requirements, the various processes may be classified as shown in table 1.1.

TABLE 1.1. Classification of Non-conventional Manufacturing Processes

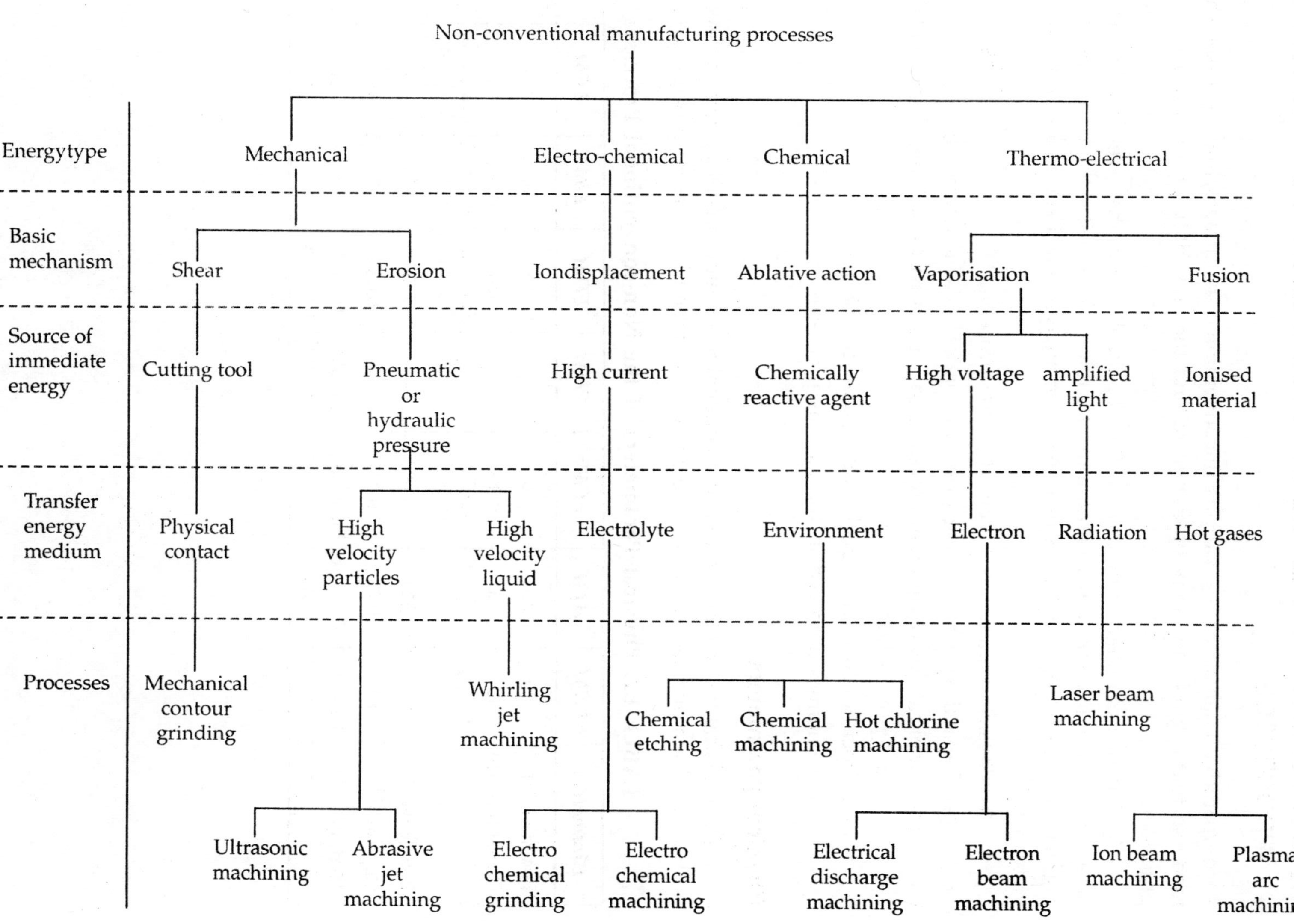

1.2 COMPARATIVE ANALYSIS OF UNCONVENTIONAL MANUFACTURING PROCESSES

A comparative analysis of the various unconventional manufacturing processes should be made so that a guide-line may be drawn to find the suitability of application of different processes.

A particular manufacturing process found suitable under the given conditions may not be equally efficient under other conditions. Therefore, a careful selection of the process for a given manufacturing problem is essential. The analysis has been made from the point of view of :

(*i*) Physical parameters involved in the processes;

(*ii*) Capability of machining different shapes of work material;

(*iii*) Applicability of different processes to various types of material, *e.g.* metals, alloys and non-metals;

(*iv*) Operational characteristics of manufacturing and

(*v*) Economics involved in the various processes.

Physical parameters

The physical parameters of non-conventional machining processes have a direct impact on the metal removal as well as on the energy consumed in different processes. (Table 1.2)

TABLE 1.2. Physical Parameters of the Non-conventional Processes

Parameters	*USM*	*AJM*	*ECM*	*CHM*	*EDM*	*EBM*	*LBM*	*PAM*
Potential (V)	220	220	10	—	45	150000	4500	100
Current (Amp)	12 (A.C.)	1.0	10000 (D.C.)	—	50 (Pulsed D.C.)	0.001 (Pulsed D.C.)	2 (Average 200 Peak)	500 (D.C.)
Power (W)	2400	220	100000	—	2700	150	—	50000
Gap (m.m.)	0.25	0.75	0.20	—	0.025	100	150	7.5
Medium	Abrasive in water	Abrasive in gas	Electrolyte	Liquid chemical	Liquid dielectric	Vaccum	Air	Argon or hydrogen

From a comparative study of the effect of metal removal rate on the power consumed by various non-conventional machining processes shown in fig. 1.2.

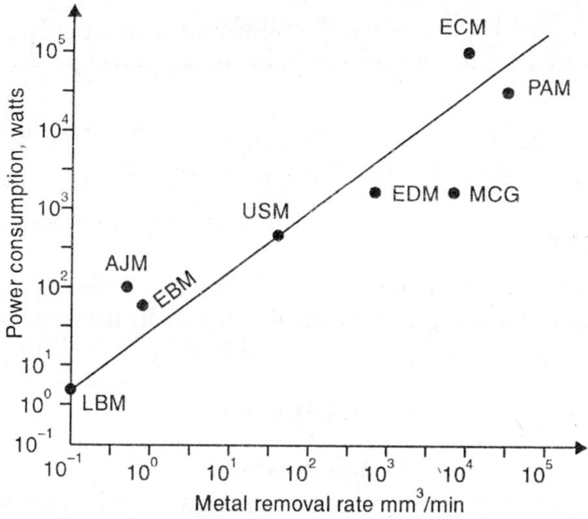

Fig. 1.2. *Effect of metal removal rate on power consumption.*

It is found that some of the processes (*e.g.* EBM, ECM) above the mean power consumption line consume a greater amount of power than the processes (*e.g.* EDM, PAM, ECG) below the mean power consumption line. Thus, the capital cost involved in the processes (EBM, ECM etc.) lying above the mean line is high whereas for the processes below that line (*e.g.*, EDM, PAM, MCG) is comparatively low.

Capability to shape

The capability of different processes can be analysed on the basis of various machining operation point of view such as micro-drilling, drilling, cavity sinking, pocketing (shallow and deep), contouring a surface, through cutting (shallow and deep) etc.

TABLE 1.3. Shape Application of Non-conventional Processes

	Holes				*Trough cavities*		*Surfacing*		*Trough cutting*	
Process	***Precision small holes***		***Standard***		***Precision***	***standard***	***Double contouring***	***Surface of revolution***	***Shallow***	***deep***
	Dia < .025 mm	***Dia > .025 mm***	***Length < 20 mm***	***Length > 20 mm***						
USM	—	—	good	poor	good	good	poor	—	poor	—
AJM	—	—	fair	poor	poor	fair	—	—	good	—
ECM	—	—	good	good	fair	good	good	fair	good	good
CHM	fair	fair	—	—	poor	fair	—	—	good	—
EDM	—	—	good	fair	good	good	fair	—	poor	—
LBM	good	good	fair	poor	poor	poor	—	—	good	fair
PAM	—	—	fair	—	poor	poor	—	poor	good	good

For micro-drilling operation, the only process which has good capability to micro-drill is laser beam machining while for drilling shapes having slenderness ratio, $\frac{l}{D}$ < 20, the process USM, ECM and EDM will be most suitable.

EDM and ECM processes have good capability to make pocketing operation (shallow or deep).

For surface contouring operation, ECM process is most suitable but other processes except EDM have no application for contouring operation.

Applicability to materials

Materials applications of the various machining methods are summarised in the table 1.4 and table 1.5. For the machining of electrically non-conducting materials, both ECM and EDM are unsuitable, whereas the mechanical methods can achieve the desired results.

TABLE 1.4

	Metals Alloys				
Process	***Aluminium***	***Steel***	***Super alloy***	***Titanium***	***Refractory material***
USM	Poor	Fair	Poor	Fair	Good
AJM	Fair	Fair	Good	Fair	Good
ECM	Fair	Good	Good	Fair	Fair
CHM	Good	Good	Fair	Fair	Poor
EDM	Fair	Good	Good	Good	Good
EBM	Fair	Fair	Fair	Fair	Good
LBM	Fair	Fair	Fair	Fair	Poor
PAM	Good	Good	Good	Fair	Poor

USM is suitable for machining of refractory type of material while AJM are for super alloys and refractory materials.

TABLE 1.5

	Non-Metals		
Process	***Ceramics***	***Plastic***	***Glass***
USM	Good	Fair	Good
AJM	Good	Fair	Good
ECM	—	—	—
CHM	Poor	Poor	Fair
EDM	—	—	—
EBM	Good	Fair	Fair
LBM	Good	Fair	Fair
PAM	—	Poor	—

Machining characteristics

The machining characteristics of different non-conventional processes can be analysed with respect to :

(*i*) Metal removal rate
(*ii*) Tolerance maintained
(*iii*) Surface finish obtained
(*iv*) Depth of surface damage
(*v*) Power required for machining

The process capabilities of non-conventional manufacturing processes have been compared in table 1.6.

The metal removal rates by ECM and PAM are respectively one-fourth and 1.25 times that of conventional whereas others are only a small fractions of it.

Power requirement of ECM and PAM is also very high when compared with other non-conventional machining processes. This involves higher capital cost for those processes. ECM has very low tool wear rate but it has certain fairly serious problems regarding the contamination of the electrolyte used and the corrosion of machine parts.

The surface finish and tolerance obtained by various processes except PAM is satisfactory.

TABLE 1.6

Process	*MRR (mm^3/min)*	*Tolerance (μ)*	*Surface (μ) CLA*	*Depth of surface damage (μ)*	*Power (watts)*
USM	300	7.5	0.2–0.5	25	2400
AJM	0.8	50	0.5–1.2	2.5	250
ECM	15000	50	0.1–2.5	5.0	100000
CHM	15	50	0.5–2.5	50	—
EDM	800	15	0.2–1.2	125	2700
EBM	1.6	25	0.5–2.5	250	150 (average) 2000 (peak)
LBM	0.1	25	0.5–1.2	125	2 (average)
PAM	75000	125	Rough	500	50000
Conventional machining	50000	50	0.5–5.0	25	3000

Economics of the processes

The economics of the various processes are analysed on the basis of following factors and given in Table 1.7.

(*i*) Capital cost
(*ii*) Tooling cost
(*iii*) Consumed power cost
(*iv*) Metal removal rate efficiency
(*v*) Tool wear.

TABLE 1.7

Process	*Capital cost*	*Tooling cost*	*Power consumption cost*	*Material removal rate efficiency*	*Tool wear*
USM	L	L	L	H	M
AJM	VL	L	L	H	L
ECM	VH	M	M	L	VL
CHM	M	L	H*	M	VL
EDM	M	H	L	H	H
EBM	H	L	L	VH	VL
LBM	L	L	VL	VH	VL
PAM	VL	L	VL	VL	VL
MCG	L	L	L	VL	L

* indicates cost of chemicals.

The capital cost of ECM is very high when compared with traditional mechanical contour grinding and other non-conventional machining processes whereas capital costs for AJM and PAM are comparatively low.

EDM has got higher tooling cost than other machining processes.

Power consumption is very low for PAM and LBM processes whereas it is greater in case of ECM.

The metal removal efficiency is very high for EBM and LBM than for other processes.

In conclusion, the suitability of application of any of the processes is dependent upon various factors and must be considered all or some of them before applying non-conventional processes.

REVIEW QUESTIONS

1. Justify the need of unconventional manufacturing process in today's industries.
2. What are the basic limitations of conventional manufacturing process ? Explain.
3. What are the basic factors upon which the unconventional manufacturing processes are classified ? Explain.
4. List five conventional and five related unconventional manufacturing processes used in industries.
5. Distinguish between conventional and unconventional manufacturing processes.
6. Why are the unconventional manufacturing processes not completely taking over the conventional manufacturing processes ? Explain.
7. Unconventional machining processes yield low rates of material removal compared to conventional processes even then they have gained wide popularity. Discuss why ?
8. Enumerate the limitations of conventional manufacturing processes particularly in the light of present day manufacturing environment.
9. ECM, EDM, USM etc. are commonly referred to as unconventional machining processes, what is unconventional in these processes ? Explain.

❑❑❑

Abrasive Jet Machining

2.0 MECHANICAL PROCESSES

The material removal by mechanical processes involve shear, erosion and abrasion. The shear mechanism involves a cutting tool as an immediate source of energy and the machining operation is performed by direct physical contact of the tool with the workpiece.

Erosion or abrasion of work material are created by Pneumatic or hydraulic pressure by the flowing high velocity abrasive particles or fluid. Erosion or abrasion also take place by the rapid and repetitive impingement of a tool on the work surface in the midst of a flowing abrasive slurry. The mechanical processes employing above mechanism of material removal are :

(*i*) Abrasive jet machining
(*ii*) Ultrasonic machining
(*iii*) Water jet machining

The various aspects of these three mechanical processes have been analysed here.

2.1 PROCESS PRINCIPLES

Abrasive Jet Machining (AJM) removes material *through* the action of a *focussed* stream of abrasive-laden gas. Micro-abrasive particles are propelled by inert gas at velocities of upto 300 m/sec. When directed at a workpiece, the *resulting* erosion can be used for cutting, etching, cleaning, deburring, polishing and drilling. The basic scheme of abrasive jet machining process is shown in fig. 2.1.

Material removal occurs through a chipping action, which is especially effective on hard, brittle materials such as glass, silicon, tungsten, and ceramics. Soft, resilient materials, such as rubber and some plastics, resist the chipping action and thus are not effectively processed by AJM. No workpiece chatter or vibration occurs with this process because the large quantity and small mass of the abrasives result in uniform loading of the part. This further enables AJM to produce fine, intricate detail in extremely brittle objects.

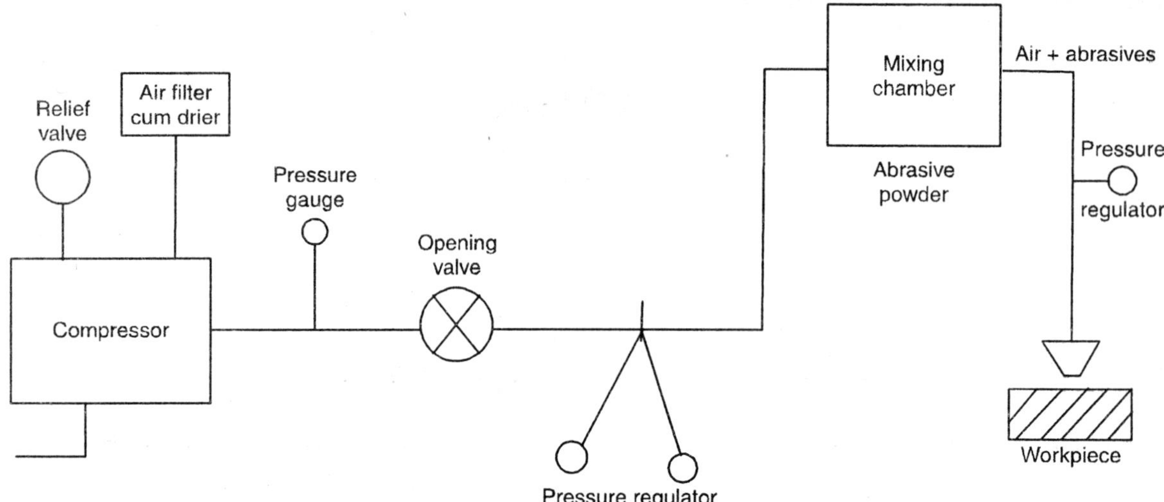

Fig. 2.1. *Scheme of abrasive jet machining*

The inside diameter of the nozzle used in the process ranges from 0.075 to 0.4 mm whereas the exit velocity of the abrasives from the nozzle is maintained between 200 and 400 m/sec. The distance of the nozzle tip from the work surface at the time of machining is known as "stand-off distance" which varies from 0.7 to 1.0 mm. The size of abrasive particle is usually taken as 10 to 50 microns.

2.2 ABRASIVE JET MACHINING SET-UP

Gas propulsion system

This supplies clean and dry gas (air, nitrogen or CO_2) to propel the abrasive particles. The gas may be supplied either by a compressor or a cylinder. In case of a compressor, air filter-cum-drier should be used to avoid water or oil contamination of the abrasive powder. The gas should be nontoxic, cheap and easily available.

Abrasive feeder

The abrasive feeder controls the abrasive quantity by inducing vibration to the feeder. The particles are propelled by carrier gas to a mixing chamber. The air-abrasive mixture moves further to the nozzle. The nozzle imparts high velocity to the mixture which is directed at the workpiece surface. The material removal occurs due to the erosive action of the jet of air-abrasive mixture impinging on the workpiece surface.

Machining chamber

The machining chamber is well closed so that the concentration of the abrasive particles around the working chamber does not reach to the harmful limit. Machining chamber is equipped with a vacuum dust collector. Special consideration should be given to the dust collection system if the toxic material (say, beryllium) are being machined.

2.3 PROCESS PARAMETERS

Major AJM process variables that affect the removal rate are :

(*i*) Abrasive particles
(*ii*) Velocity of fluid
(*iii*) Carrier gas
(*iv*) Work material
(*v*) Stand-off distance
(*vi*) Nozzle design.

Abrasive particles

The abrasive particles should have irregular shape and consist of sharp edges. The composition of the abrasives depends on the type of machining process to be carried out.

The abrasives used for cutting are aluminium oxide and silicon carbide, whereas sodium bicarbonate, dolomite, glass beads etc. are used for cleaning, etching, deburringand polishing. Re-use of abrasives is not recommended because not only does its cutting ability decrease, but contamination also clogs the orifice of the nozzle.

Available standard sizes of commercial abrasive particles are usually 10, 27, 40 and 50 microns.

The material removal rate is mainly governed by the flow rate and size of the abrasive particles. This is evident from graph shown in fig. 2.2. In this case, the nozzle orifice diameter and the tip distance are 0.4 mm and 0.8 mm respectively and air is used as a carrier gas.

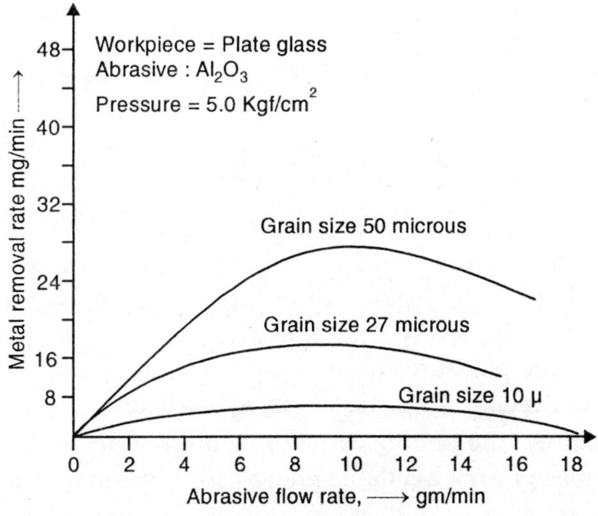

Fig. 2.2. *Effect of flow rate and size of abrasive on metal removal rate*

From the fig 2.2, it is clear that at a particular pressure the material removal rate increases with the increase of abrasive flow rate and is influenced by the size of the abrasive particles. But after reaching an optimum value, the material removal rate decreases with further increase of abrasive flow rate. This is owing to the fact that mass flow rate of the gas decreases with the increase of abrasive flow rate and hence the mixing ratio increases causing a decrease in material removal rate because of the decreasing energy available for erosion.

TABLE 2.1

Abrasive	*Grain size*	*Applications*
Aluminium oxide (Al_2O_3)	12, 20, 25 μ	Cutting & grooving
Silicon Carbide (SiC)	25, 40 μ	Cutting & grooving
Sodium bicarbonate	27 μ	Light finishing below 50°C
Dolomite	200 mesh	Etching and polishing
Glass beads	0.635 to 1.27 mm	Light polishing and fine deburring

Velocity of fluid

The kinetic energy of the abrasive jet is utilized for metal removal by erosion. The velocity of the carrier gas conveying the abrasive particles changes considerably with the change of abrasive particle density.

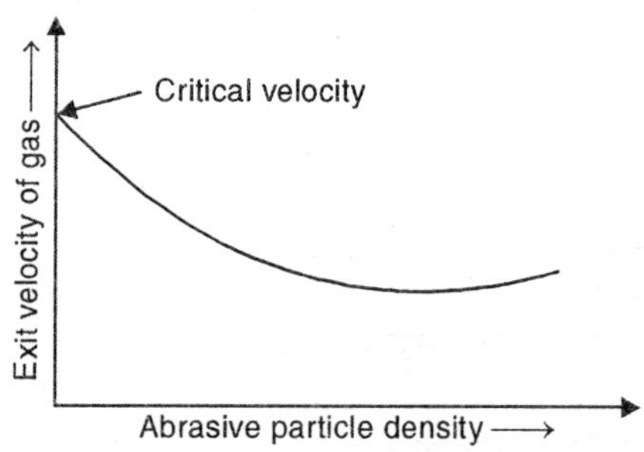

Fig. 2.3. *Effect of abrasive particle density on gas exit velocity.*

The exit velocity of the gas can be increased to critical velocity when the internal gas pressure is nearly twice the pressure at the exit of the nozzle for an abrasive particle density of zero. If the density of the abrasive particles is gradually increased, the exit velocity will go on decreasing for the same pressure condition. It is due to the fact that the kinetic energy of the fluid is utilized for transporting the abrasive particles.

The increased mass flow rate of abrasive will result in a decreased velocity of fluid and will thereby cause a decrease in the available energy for erosion and ultimately the material removal rate.

Mixing ratio is defined as :

$$\text{Mixing Ratio (MR)} = \frac{\text{Mass flow rate of abrasive}}{\text{Mass flow rate of fluid or gas}}$$

A large value of MR should result in higher rates of metal removal but a large abrasive flow rate has been found to adversely influence jet velocity.

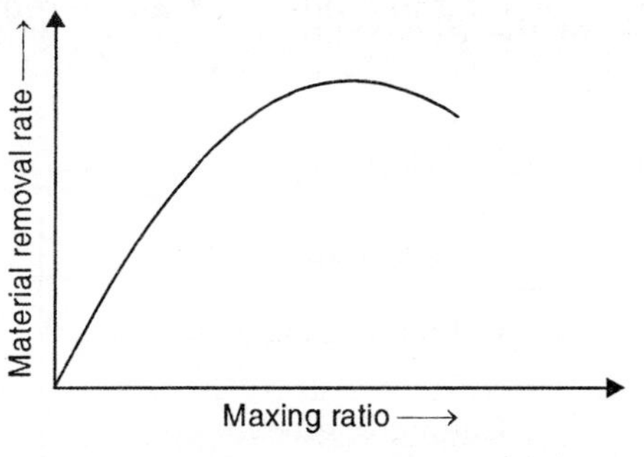

Fig. 2.4. *Effect of mixing ratio on material removal rate*

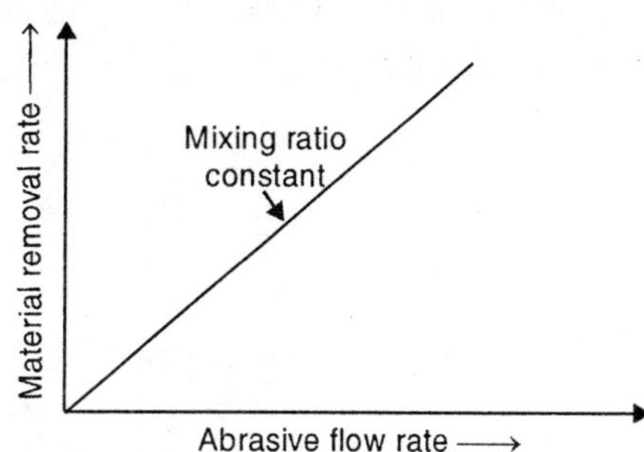

Fig. 2.5. *Effect of abrasive flow rate on metal removal rate*

The material removal rate can be improved by increasing the abrasive flow rate provided the mixing ratio can be kept constant. The mixing ratio can be kept unchanged only by a simultaneous increase of both gas and abrasive mass flow. The abrasive flow rate can be increased by increasing the mass flow rate of the carrier gas or fluid. This is only possible by increasing the internal gas pressure as shown in fig. 2.6.

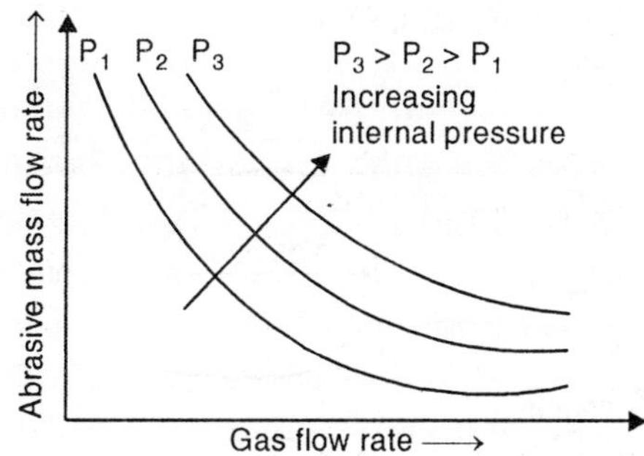

Fig. 2.6. *Effect of gas flow rate on abrasive flow rate with increasing pressure.*

As a matter of fact, the material removal rate will increase with the increase of gas pressure.

Carrier gas

The carrier gas of propellant is used in AJM process is usually air, carbon dioxide (CO_2) and nitrogen (N_2) but never oxygen. The carrier gas must not flare excessively when discharged from the nozzle into the atmosphere. Further, the gas should be nontoxic, cheap, easily available and capable of being dried and cleaned without difficulty.

Work material

AJM is suitable for processing of brittle materials, such as glass, ceramics, refractories etc. Most of the ductile materials are practically unmachinable by AJM.

Stand-off distance

Stand-off distance is defined as the distance between the face of the nozzle and the working surface of the work. Stand-off distance has a considerable effect on metal removal rate as well as accuracy.

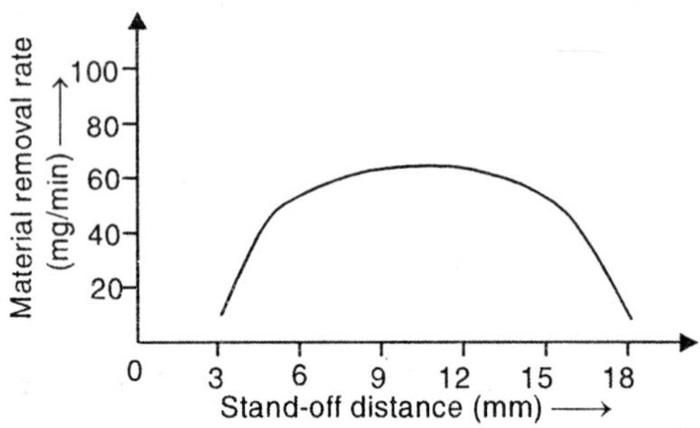

Fig. 2.7

From the figure 2.7, it is clear that material removal rate increases with the increase of tip distance upto a certain limit after which it remains unchanged for a certain tip distance and then falls gradually.

The stand-off distance has also a direct effect on the width of cut owing to the outward flaring of the abrasive jet after a short distance.

This is made clear in fig. 2.8.

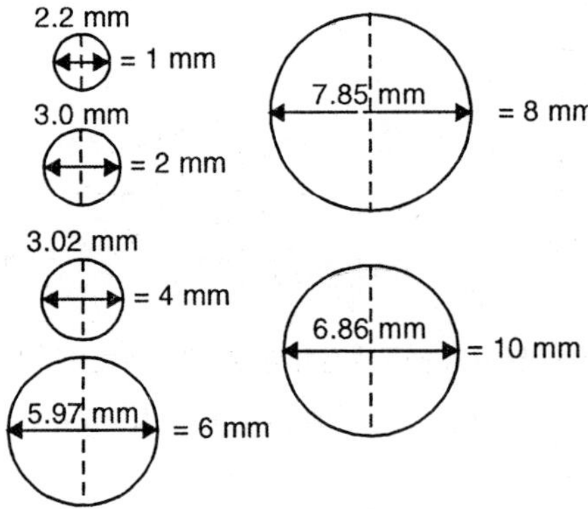

Fig. 2.8

Nozzle design

The nozzle has to withstand the erosive action of abrasive particles and hence, must be made of materials that can provide high resistance to wear. In general, nozzles are made of tungston carbide (WC) or sapphire having regular orifices or slots depending on the utility.

The dimensions of various kinds of slots and the nozzle life are indicated in the table 2.2 for various nozzle materials.

TABLE 2.2

Nozzle material	*Round shape nozzle dia. mm*	*Rectangular shape slot, dimension, mm*	*Life of nozzle, hours*
Tungsten carbide (WC)	0.2 to 1.0	0.075 × 0.5 to 0.15 × 2.5	12 to 30
Sapphire	0.2 to 0.8		300

2.4 PROCESS CAPABILITIES

Material removal rates with AJM are considered low, typically only 0.016 cm^3/min. However the ability to produce intricate detail in hard, brittle materials makes up for the low removal rates. Slots as narrow as 0.12 – 0.25 mm can be produced when stray cutting is minimized with rectangular nozzles. Tolerances of ± 0.12 mm are easily obtained, and ± 0.05 mm can be achieved with proper techniques. The minimum radius that can be produced is 0.2 mm. Surface finish range from 10 to 50 μin.

Steel as thick as 1.5 mm and glass 6.3 mm thick have been cut by AJM but at very slow rates and with large amounts of taper.

2.5 APPLICATION EXAMPLES

AJM has been successfully employed to manufacture small electronic devices consisting of a 0.38 mm thick wafer of silicon brazed to a 0.75 mm thick tungston disk. After two materials are brazed together, the silicon wafer must be trimmed and beveled without harming the tungston disk. To accomplish this task, an AJM nozzle is mounted at the desired angle and directed at the slowly rotating part. With this technique, the unwanted silicon is trimmed off each part in less than 1 minute.

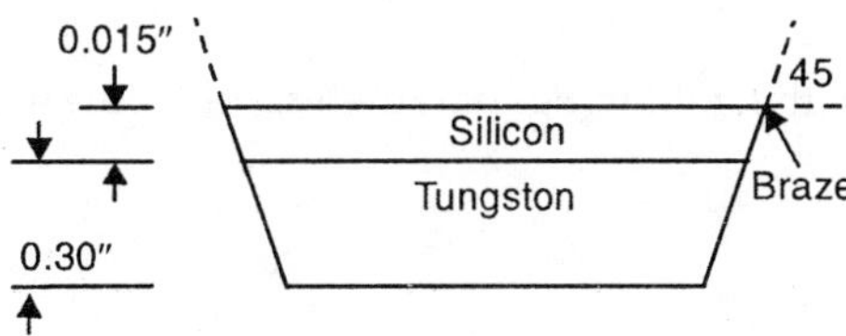

Fig. 2.9. *AJM trimmed electronic component*

For deburring purpose, the AJM is used in manufacture of small biomedical analysis packages. Small plastic cubes are cross-drilled with two 0.34 mm diameters holes. Tiny burrs are created internally at the inter-section of the two holes and must be removed without producing scratches.

A variety of other diverse jobs have been successfully performed by AJM. These include cutting threads into glass rods, deflashing small castings, die and mold touch up, cutting titanium foil and drilling glass wafers.

2.6 PROCESS SUMMARY

Advantages

(*i*) Machines heat sensitive and fragile materials.
(*ii*) Very low capital cost.
(*iii*) No part chatter or vibration.
(*iv*) Good for difficult to reach areas.
(*v*) Machines very hard materials.

Disadvantages

(*i*) Stray cutting.
(*ii*) Low removal rate.
(*iii*) Taper.
(*iv*) Short nozzle stand-off when used for cutting.
(*v*) Paticles can embed into workpiece.
(*vi*) Lack of dust removal system.

REVIEW QUESTIONS

1. Explain the basic working principle of Abrasive Jet Machining (AJM) process with suitable sketches.
2. Write the applications of different types of abrasives used in AJM. Also, comment on accuracy machining of AJM.
3. Discuss the mechanism of material removal for abrasive jet machining. Explain how would you select the best possible abrasive and the nozzle material to be used in this process.
4. Discuss why the AJM technique, when applied to ductile materials, leads to a low rate of metal removal.
5. Discuss the effects of the following parameters on working accuracy and rate of metal removal in AJM :
   (*i*) Grain size.
   (*ii*) Jet velocity.
   (*iii*) Stand off distance.
6. Describe at least three typical engineering applications of AJM.
7. Write the applications of different types of abrasives used in AJM. Also comment on accuracy machining of AJM.

❑❑❑

Ultrasonic Machining (USM)

3.0 INTRODUCTION

Ultrasonic machining (USM) is a mechanical material removal process used to erode holes and cavities in hard or brittle workpieces by using shaped tools, high frequency mechanical motion, and an abrasive slurry. Being non chemical and non-thermal process, materials are not altered either chemically or metallurgically. Ultrasonic machining is able to effectively machine all materials harder than HRC 40, whether or not the material is an electrical conductor or an insulator.

In 1950, first time Lewis Balamuth had discovered the ability of an ultrasonically vibrating tool to interact with an abrasive mixture and produce material removal.

The USM process begins with the conversion of low frequency electrical energy to a high-frequency electrical signal, which is then fed to a transducer. The transducer is a device that converts the high-frequency electrical signal to high-frequency linear mechanical motion. For conversion of electrical energy into mechanical vibrations, the piezoelectric effect in natrural or synthetic crystals or the magnetostrictive effect exhibited by some metals is utilized.

3.1 FUNDAMENTAL PRINCIPLES

In USM process, ultrasonic waves or vibrations are transformed by means of a magnetostrictive transducer into mechanical vibrations of small amplitude and high frequency. A horn (stepped or exponential) in attached to the end of the transducer for mechanical amplification of the amplitude of vibration of the transducer and a tool of any desired shape is attached at the free end of the horn to serve the purpose of cutting. The ultrasonic cutting operation is schematically shown in figure 3.1.

As the tool vibrates with a specific frequency, an abrasive slurry (usu ally a mixture of abrasive grains and water of definite proportion) is made to flow through the tool-work piece interface. The impact force arising out of the vibration of the tool end and the flow of slurry through the work-tool interface actually causes thousands of microscopic abrasive grains to remove the work material by abrasion.

The ultrasonic vibrations imparted to the fluid medium surrounding the tool have three-fold action as follows :

(*i*) To bring about the ultrasonic dispersion effect rapidly in the machining fluid medium between the tool end and the machining surface of the workpiece.

(*ii*) To cause violent circulation of the fluid as a result of ultrasonic micro-agitation.

(*iii*) To cause the cavitation effect in the fluid medium arising out of the ultrasonic vibration of the tool in the fluid medium.

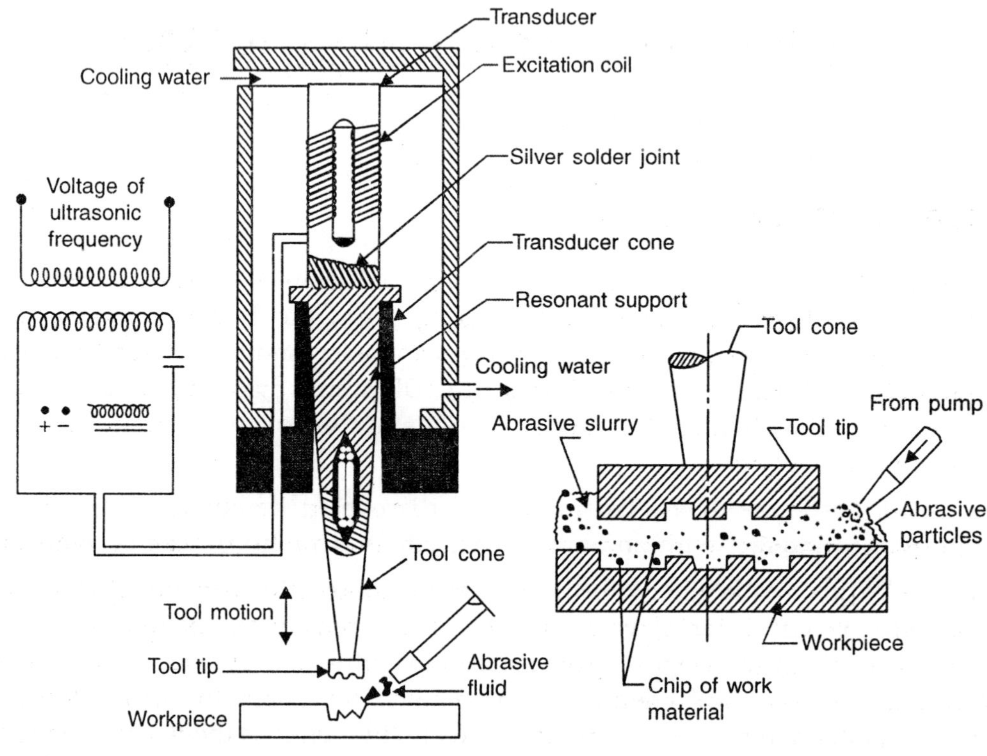

Fig. 3.1

The theoretical analysis reveals that USM is a form of abrasion and material is removed by :

(*i*) shear deformation;

(*ii*) brittle fracture of work material by impact.

(*iii*) cavitation

(*iv*) chemical reaction.

3.2 EQUIPMENTS

The machines used for USM range from small to large capacity machine tools. In addition to the part-size capacity of a USM machine, suitability for a particular application is also determined by the power rating. The power of a USM machine is rated in watts and can range from 40 W to 2400 W.

All USM machines share common subsystems regardless of the physical size or power. The most important of these subsystems are the power supply, tool cone and tool tip, abrasive slurry, tool feeding mechanism and transducers.

Power supply

The power supply used for USM is more accurately characterised as a high power sine wave generator that offers the user control over both the frequency and power of the generated signal. It converts low frequency (60 Hz) electrical power to high frequency (approx. 20 KHz) electrical power. This electrical signal is then supplied to the transducer for conversion into mechanical motion.

Tool cone and tool tip

Tool holder holds and connects the tool to the transducer. It transmits the energy and amplifies the amplitude of vibration.

Commonly used materials for the tool holder are monel, titanium and stainless steel. Beacuse of good brazing as well as acoustic properties, monel is commonly used material for low amplitude application. High amplitude application requires good fatigue strength of tool holder material. Futhter, tool holder may be amplifying or non-ampilfying. Non-amplifying tool holders have circular cross-section and give the same amplitude at both the ends, *i.e.* input end and output end. Amplifying tool holders give as much as six times increased tool motion. It is achieved by stretching and relaxing the tool-holder material. Such a tool-holder yields MRR up to 10 times higher than non-amplifying tool holder. However, amplifying tool holders are more expensive, demand higher operating cost and yield poorer surface quality.

Tools are usually made of relatively ductile materials (brass, stainless steel, mild steel etc.) so that the tool wear rate (TWR) can be minimized. Value of the ratio of TWR and MRR depends upon the kind of abrasives, workpiece material and tool material. Surface finish of the tool is important because it will affect the surface finish obtained on the workpiece.To safeguard tool and tool-holder against their early fatigue failure. They should not have scratches or machining marks. Tools should be properly designed to account for overcut.

Abrasive slurry

The main purposes of using the slurry are :

(*i*) To carry abrasive to the machining zone,

(*ii*) To take away the wear particles.

(*iii*) To cool the tool and the workpiece.

The cutting element in USM is a slurry which is a mixture of water and abrasive particles. The abrasives which are generally used in practice are :

(*i*) Boron Carbide (B_4C)

(*ii*) Silicon Carbide (SiC)

(*iii*) Aluminium oxide (Al_2O_3)

The size of the abrasive particles are chosen on the basis of surface finish required. Coarser size of particles cut at a faster rate than the finer grits but the surface finish obtained is not so good as with the finer grits. Hence, to achieve the desired cutting action, it is necessary to choose the abrasive size carefully for individual job requirements. The factors which influence the selection of proper abrasives for USM are essentially the same as those governing the choice of grinding wheels for conventional grinding operations.

The factors which should be considered for proper selection of abrasive are :

(*i*) Type of material to be machined.

(*ii*) Hardness of the material.

(*iii*) Amount of material removal desired.

(*iv*) Suface finish required.

Boron carbide is the most widely used abrasive in USM for the following reasons:

(*i*) It is nearly two times harder than silicon carbide and has greater resistance to fracture.

(*ii*) It can cut at a faster rate than any other type of abrasive.

(*iii*) It has the capability of withstanding very high vibrational and impact forces encountered in the USM process.

(*iv*) Close tolerance and proper surface finish can be achieved with its use.

The liquid used in the slurry can be water, benzene, glycerol, oil etc. Of these, water is found to give the best results due to its lower viscosity. It is clear from the experimental study of Pentland that the metal removal rate decreases with increase of viscosity.

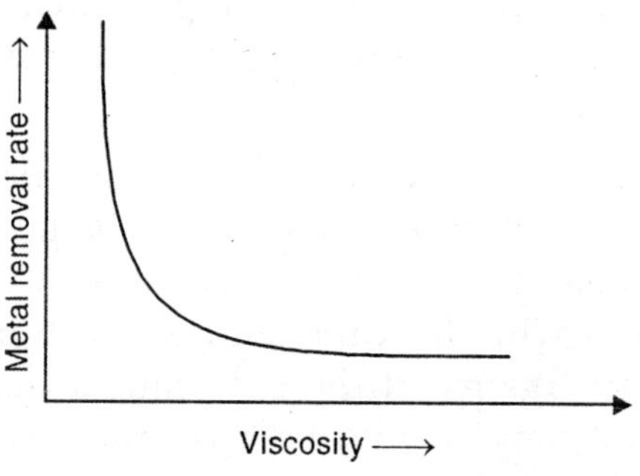

Fig. 3.2. *Effect of viscosity on metal removal rate*

Plain water is preferable. The experiment of Neppiras has shown the effect of percentage of abrasive on metal removal rate as shown in figure 3.3.

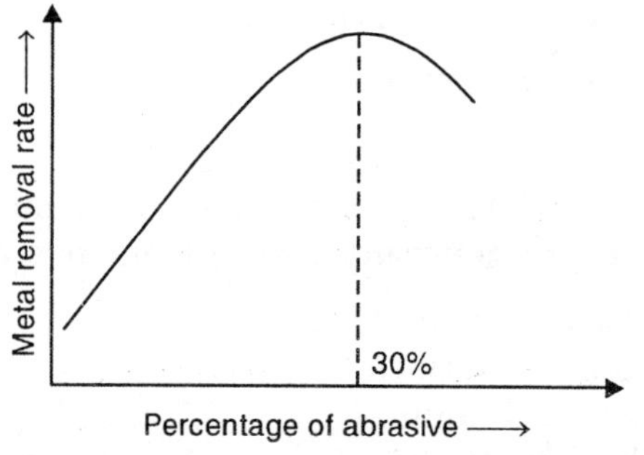

Fig. 3.3. *Effect of percentage of abrasive on m.m.r.*

Tool feeding mechanism

Feeding of the tool is provided for applying a working force between the tool and the work piece and for sustaining this force during cutting operation. Gravity feeding mechanism is generally preferred for its simplicity. In this case, the driving force is the difference between the weight of the head and that of the counter weight.

The spring load system of feeding can also be provided because of its compactness and sensitivity arising out of the small number of rubbing surfaces. But in this case feeding varies as the penetration of the tool into the workpiece.

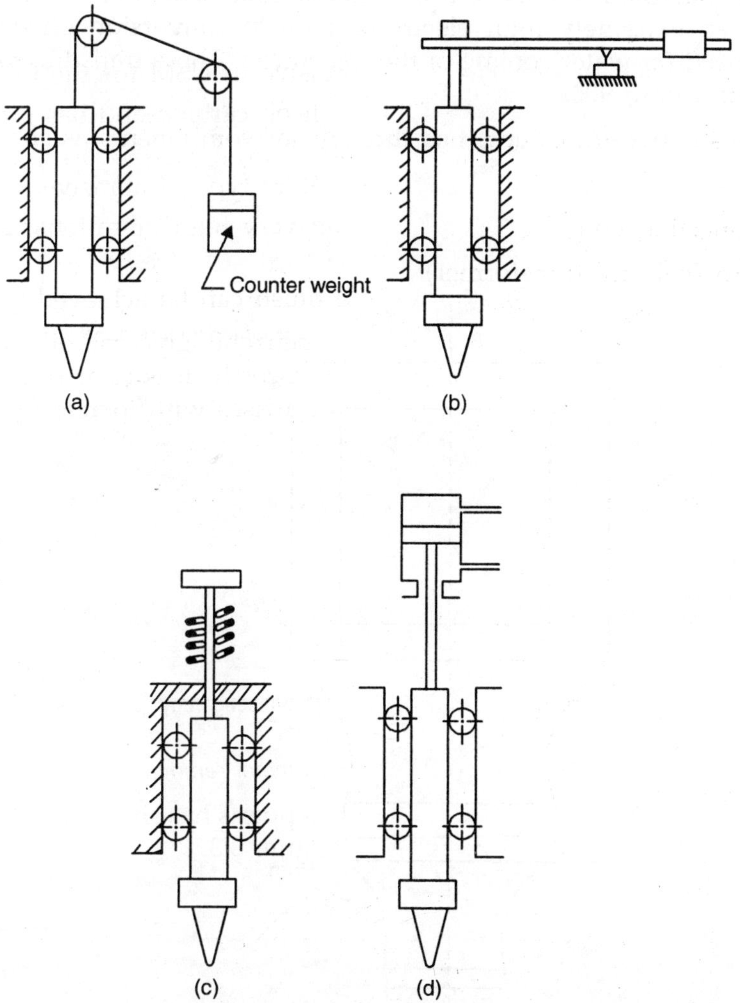

Fig. 3.4. *Tool feed systems*

All the aforesaid feeding mechanisms are reliable provided they maintain a considerable degree of accuracy and the contact surfaces have low frictional losses. Balls are often used in the guides for reducing friction. Pneumatic or hydraulic system of feeding does not yield satisfactory performance since a very small feed rate is desirable in USM.

Transducer

Transducer is the main component of an ultrasonic machine which is used for converting electrical energy into mechanical energy in the form of high frequency vibration. Usually magnetostrictive and peizo-electric transducer are chosen for smooth performance at full power for extended periods.

Piezoelectric transducers used for USM generate mechanical motion through the piezoelectric effect by which certain materials, such as quartz or lead zirconate titanate, will generate a small electric current when compressed. Conversely, when an electric current is applied to one of these materials, the material increases minutely in size. When the current is removed, the material instantly returns to its original shape. Piezoelectric transducers, by nature exhibit an extremely high electromechanical conversion efficiency (96%), which eliminates the need for water cooling of the transducer. These transducers are available with power capabilities upto 900w.

A magnetostrictive transducer has three major components as shown in figure 3.5.

(*i*) Stack

(*ii*) Connecting body

(*iii*) Horn (velocity transformer)

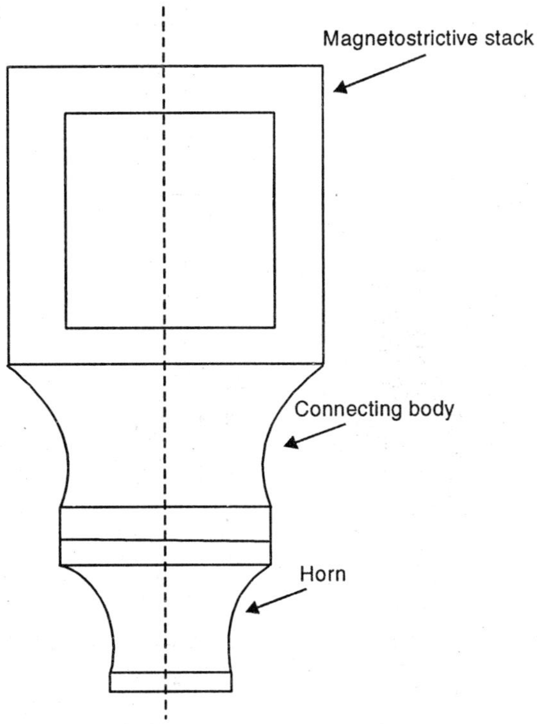

Fig. 3.5

Magnetostrictive transducers are generally employed for the machining of materials and are made of either solid or laminated magnetostrictive materials. Transducer is built with laminated sheets in preference to a solid body construction since the eddy current loss will be reduced to a large extent by using laminated construction.

3.3 MAGNETOSTRICTION

Magnetostriction means a change in the dimensions occurring in ferro-magnetic materials subjects to a periodic magnetic field. When the transducer is magnestised, a change, in length occurs. The connecting body attached to the transducer receives and transmits this change in length and it is further amplified by a horn (usually stepped or exponential type).

In an atomic system the fundamental element is the orbital electron. In many of the systems, the magnetic effects nearly neutralize each other except in the case of ferromagnetic materials. In these cases, an exchange force exists which causes the magnetic fields within a volume of 10^{-8} cm^3 to lie parallel to one another. These domains of magnetic moments can be aligned by an external magnetic field. Each domain is magnetised to saturation point, but depending on the crystal structure of the material, the magnetic field may be along any one of a number of fixed directions known as directions of easy magnetisation.

With increasing external magnetic field, certain domains grow in size, taking over the other differently oriented domains until the complete crystal becomes one large domain. During this process, the material expands or contracts externally in different directions with different degrees. All magnetostrictive materials are sensitive to temperature. The materials lose their magnetic properties with rise in temperature up to curie temperature where the magnetic property is found to vanish.

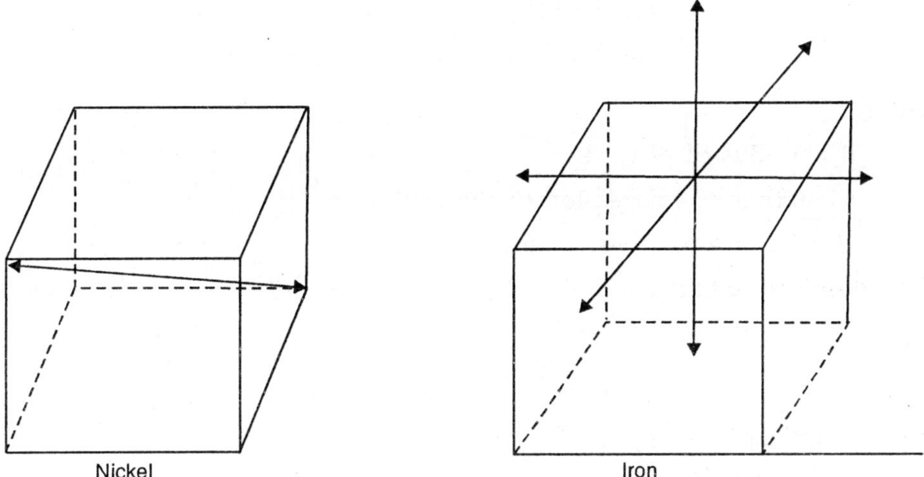

Fig. 3.6. *Growth of material in different directions*

The materials which are capable of producing and withstanding ultrasonic waves are the alloys of nickel, chromium, iron, cobalt invar, monel (68% Ni, 28% Cu) etc. The most commonly used materials are Ni, Permalloy (45% Ni, 55% Fe) and Permedur (49% Co, 49% Fe,2% vanadium).

3.4 MECHANICS OF CUTTING

The mechanics of material removal from a surface by forcing sharp grains into it have been studied by many researchers. Based on certain assumptions, they have given estimates of the cutting rate.

Mathematical model of shaw for metal removal rate

In determining the metal removal rate, Prof Shaw and his colleagues developed a mathematical model with the following assumptions :

(*i*) All the abrasive grits are assumed to be spherical

(*ii*) Metal removal rate is based on the mechanism of fracture.

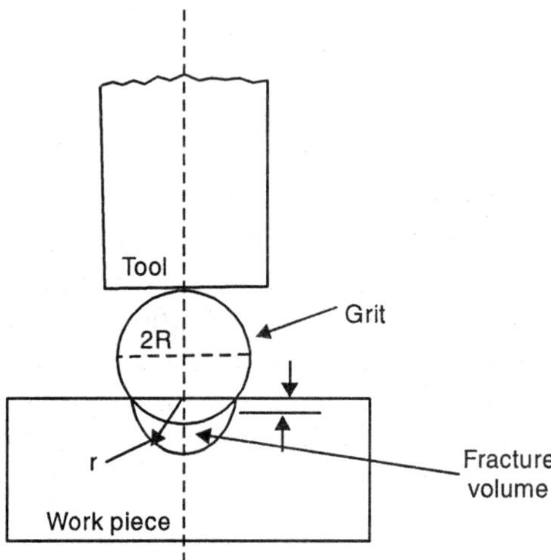

Fig. 3.7. *Scheme of Prof. Shaw's model*

Where,

R = radius of grit

r = radius of impression of the grit

H = surface hardness of the workpiece.

The volume of material removed by fracture per grit per cycle is given by

$$V = \frac{1}{2}\left(\frac{4}{3}\pi r^3\right) \quad ...(3.1)$$

From the geometry of figure 3.7

$$r^2 = R^2 - (R - \delta)^2$$

$\cong 2R\delta$ (neglecting δ^2 as δ, the depth of penetration is very small compared to R)

Therefore,

$$r = (2R\delta)^{1/2} \quad ...(3.2)$$

Thus, equation 3.1 transforms to

$$V = \frac{2}{3}\pi\ (2R\delta)^{3/2}\ \text{mm}^3/\text{grit}/\text{cycle} \quad ...(3.3)$$

where R and δ are in millimetre.

All the grits do not strike the workpiece. The smaller grits become inactive as they do not touch both the workpiece and the tool. The number of active grits can be estimated as described hereinafter.

The packing density ratio, η, has been given by

$$\eta = \frac{\text{Number of grits per area}}{\text{Maximum number of grits per area}}$$

But, closed possible density $= \dfrac{3}{\pi(2R)^2} \cong \dfrac{1}{4R^2}$

Therefore,

$$\text{Total grits per area} = \frac{\eta}{4R^2} \quad \text{...(3.4)}$$

Let the number of active grits be N.

Considering the two cases as shown in figure 3.8, it is seen that $\dfrac{\delta}{2R}$ depends mainly on the depth of impression made by each grit. When the depth of penetration is large, the number of active grits will be more and will be proportional to $\dfrac{\delta}{2R}$.

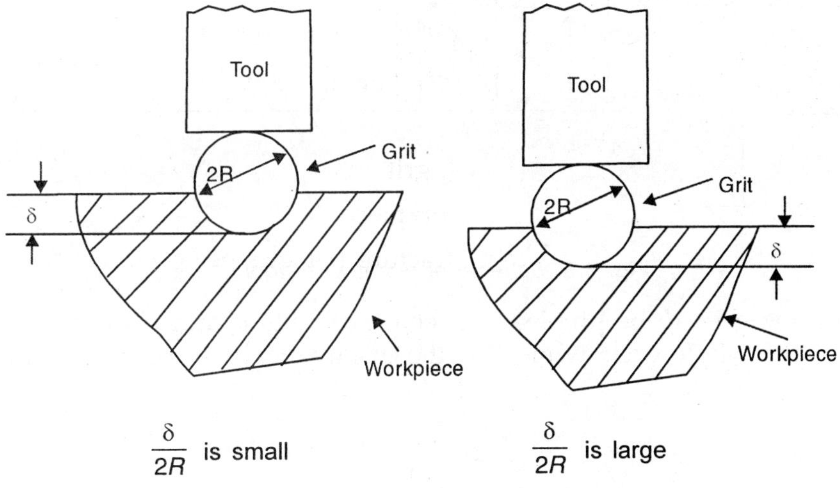

Fig. 3.8. *Variation of depth of penetration*

Thus,

$$N \propto \left(\frac{\delta}{2R}\right) \times \text{total number of grits per area}$$

or,

$$N \propto \left(\frac{\delta}{2R}\right) \times \left(\frac{\eta}{4R^2}\right)$$

hence,

$$N = \tau\left(\frac{\delta}{2R}\right) \times \left(\frac{\eta}{4R^2}\right)$$

where, τ is an arbitrary constant.

If P = average force per grit, then

$$P = \pi r^2 H \qquad ...(3.5)$$

where H = surface hardness of the workpiece in Kg/mm^2

Using equation 3.2

$$P = \pi (2R\delta) H = (2\pi RH) \delta = K \delta \qquad ... (3.6)$$

where, $K = 2\pi RH$ ... (3.7)

If the oscillating motion of the tool is assumed to be simple harmonic in nature

$$y = y_0 \sin \omega t = y_0 \sin\theta \qquad ... (3.8)$$

where, $\theta = \omega t$ and t = any time

The vibratory motion of the tool has been graphically shown in figure 3.9.

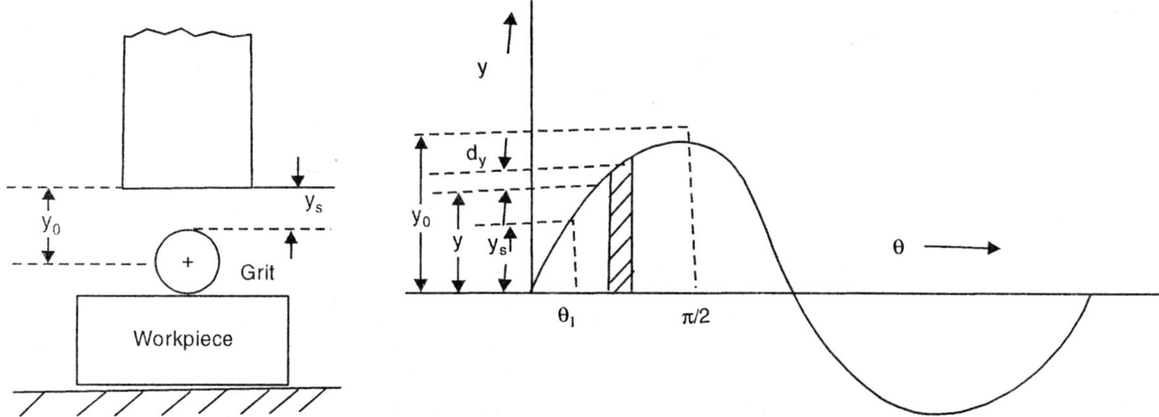

Fig. 3.9. *Scheme of vibration of tool*

Considering the grits to be pressed between the limits y_s and y_o only and P_{avg} as the average force for the whole cycle while P is the instantaneous force during impact.

$$2\pi P_{avg} = \int_{\theta_1}^{\frac{\pi}{2}} P d\theta \qquad ...(3.9)$$

where, θ_1 and $\frac{\pi}{2}$ are the angles described when the displacement of the tool is y_s and y_0 respectively.

But $P = K\delta$ and $\delta = y - y_s$

Therefore,

$$2\pi P_{avg} = \int_{\theta_1}^{\pi_2} K(y_0 \sin\theta - y_s)\, d\theta \qquad ...(3.10)$$

where, $y = y_o \sin \theta$

or, $2\pi P_{avg} = K\left[-y_0 \cos\theta - y_s\theta\right]_{\theta_1}^{\pi/2}$

$$= K\,[y_0 \cos\theta_1 - y_s(\pi/2 - \theta_1)]$$

or, $$\frac{2\pi P_{avg}}{Ky_0} = \left[\cos\theta_1 - \frac{y_s}{y_0}(\pi/2 - \theta_1)\right]$$

$$= [\cos\theta_1 - \sin\theta_1(\pi/2 - \theta_1)] \qquad \text{...(3.11)}$$

where, $$y_s = y_0 \sin\theta_1,$$

or, $$\frac{2\pi P_{avg}}{Ky_0} = f(\theta_1) \text{ or } \left(\frac{\delta}{y_0}\right) \qquad \text{...(3.12)}$$

Again, if σ be the stress developed in the tool, then

$$P_{avg} = \frac{\sigma A_r}{NA_r} = \frac{\sigma}{N} \qquad \text{...(3.13)}$$

where, A_r = actual area in action.

Besides, from the plot of $\frac{2\pi P_{avg}}{Ky_0}$ with respect to $f(\theta_1)$ or $\frac{\delta}{y_0}$ as shown in figure 3.10, it is found that

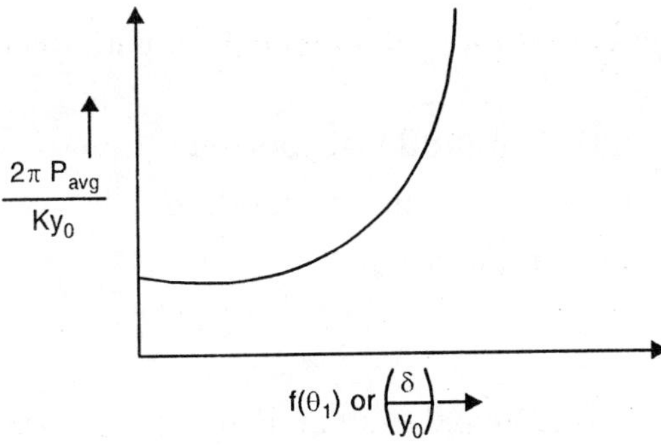

Fig. 3.10

$$\frac{2\pi P_{avg}}{Ky_0} = \left(\frac{\delta}{y_0}\right)^{3/2} \qquad \text{...(3.14)}$$

Putting the value of P_{avg} from equation 3.13 and K = 2 π R H, it is seen that,

$$\frac{2\pi P_{avg}}{Ky_0} = \frac{\sigma}{N}\frac{1}{RH}\cdot\frac{1}{y_0} = \left(\frac{\delta}{y_0}\right)^{3/2}$$

or $$\left(\frac{\delta}{y_0}\right)^{3/2} = \left(\frac{\sigma}{H}\right)\frac{1}{Ry_0}\cdot\frac{1}{N}$$

Thus, $$N = \left(\frac{y_0}{\delta}\right)^{3/2}\left(\frac{\sigma}{H}\right)\frac{1}{Ry_0} \qquad ...(3.15)$$

Metal removal rate per minute

$$V = \frac{2}{3}f\,N\pi(2R\delta)^{3/2} \qquad ...(3.16)$$

where, f = frequency of the active grits striking work surface.

Combining equation 3.15 and 3.16

$$V = \frac{2}{3}f\,\pi(2R\delta)^{3/2}\left[\frac{\sigma}{H}\left(\frac{y_0}{\delta}\right)^{3/2}\frac{1}{Ry_0}\right]$$

Thus,

$$V = 5.9fR^{1/2}{y_0}^{1/2}\left(\frac{\sigma}{H}\right) \qquad ...(3.17)$$

This is the expression for MRR obtained by Prof. Shaw which shows that $V \propto y_0^{1/2}$ and it is found to be consistent with practical results.

3.5 ANALYSIS OF PROCESS PARAMETERS

Various process parameters which influence the metal removal rate and other process criteria are :

(*i*) Effect of slurry (*ii*) Amplitude of tool vibration

(*iii*) Grit size (*iv*) Work material condition

(*v*) Power (*vi*) Frequency.

Effect of slurry

The rise in cutting rate can be achieved with an increase in slurry concentration. Neppiras further established that saturation occurs when the volume of the slurry is 30 to 40 per cent abrasive/water mixture.

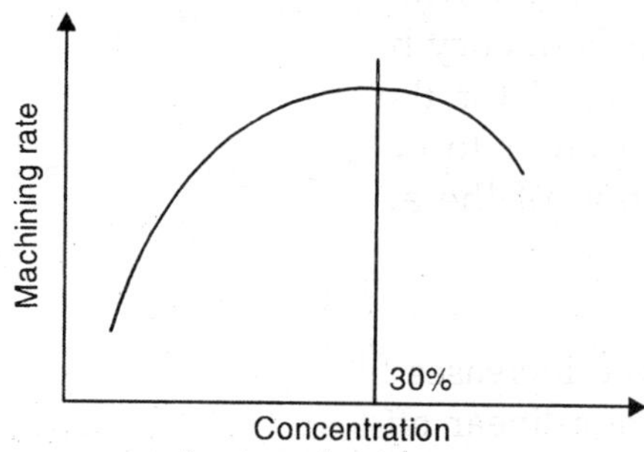

Fig. 3.11. *Effect of slurry concentration on material removal rate*

The pressure with which the slurry is fed into the cutting zone has a remarkable effect on the material removal rate.

The influence of slurry temperature on the removal rate is mainly due to cavitation effect. Cavitation is again influenced by the variation of fluid or slurry temperature. It is found that maximum metal removal rate is obtained at the temperature of about 50°C which signifies the fact that the vapour pressure of water at this temperature causes optimum cavitation and the cavitation effect is found to decrease below or above this temperature.

Amplitude of tool vibration

The tool vibration amplitude has a predominant influence on the metal removal rate in USM. With the increase in the amplitude of tool vibration, the velocity of the tool tip also increases in the same proportion. This results in a corresponding increase in the momentum of the abrasive particles. Hence, the kinetic energy of the particles will also increase as the square of the velocity and the metal removal rate will increase. The experimental results also agree with this.

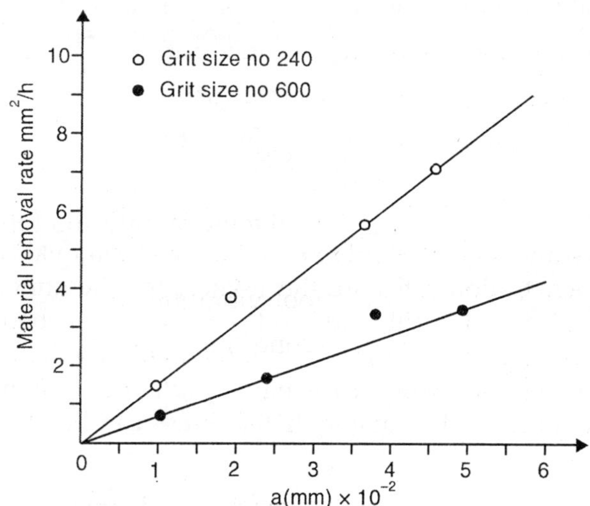

Fig. 3.12. *Relation between Amplitude of vibration and material removal rate*

Markov has shown that material removal rate is directly proportional to the particle velocity, that is amplitude/resonant frequency bears a linear relationship to machining rate. It thus implies that the frequeney used for the machining process must be the resonant frequency of the acoustic system in order to obtain greatest amplitude at the tool tip and thus achieve the maximum utilization of the acoustic system.

Grit size

Goetze claimed that cutting rate increases linearly with grain size, but the findings of Neppiras and Foskett indicated a non-linear effect of grain diameter on the removal rate. These is a limit to the effect of grain size on the rate as a very coarse powder may even cause a fall in rate.

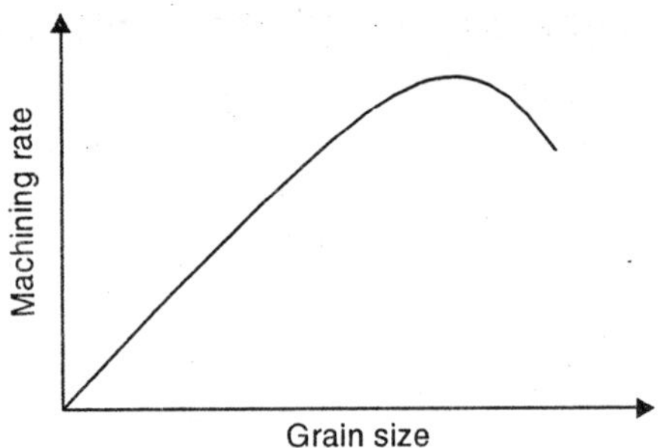

Fig. 3.13. *Relation between abrasive grain size and metal removal rate*

The optimum size is, however, governed by the amplitude of tool vibration. Surface finish is greatly influenced by the grain size.

Grain size determines the accuracy of the cavity configuration in ultrasonic machining. The hole cut is larger than the tool owing to the flow of abrasive along the sides of the hole to the bottom face of the tool. For accuracy and good surface finish, it is better to use a set of tools and more than one size of abrasive grits in stages, as follows:

Stage 1. undersize tool, high frequency, coarse grit

Stage 2. undersize tool, high frequency, finer grit

Stage 3. full size tool, low frequency, very fine grit

Work material condition

The work material hardness greatly influences the metal removal rate. In USM process, it is found that the tool penetration rates are the highest for the annealed condition, with lower rates for the normalised and still lower rates for oil quenched specimens. Similar references have also shown that the process is more suitable for hard materials than soft materials. But, experiments on brass workpieces have resulted in a high metal removal rate, thus confirming that some of the softer materials like brass can be effectively machined by USM.

Power

Power primarily determines the mass of the tool—tool holder combination that can be utilised for an application and also determines the frontal-cutting area of the tool. The more power available in a USM machine, the larger the frontal-cutting area that can be supported. Therefore, when drilling large size holes, machine power requirements can be reduced by trepanning the shape with a hollow tool rather than by using a solid tool.

Frequency

For any given tool-toolholder combination there is only one frequency at which the combination will be in resonance for efficient cutting. The higher the resonance frequency, the higher the rate of material removal.

Because the material removal rate is drastically increased when the power generator is tuned to the proper resonant frequency, it is important that the machine operator monitor the process and make frequency tuning adjustments when necessary. Tuning is required when sufficient tool material is eroded to modify the weight and therefore the resonant frequency of the tool-tool holder combination.

3.6 APPLICATION EXAMPLES

Most successful USM applications involve drilling hole, or making cavities in non-conductive ceramic materials. The selection of USM as the preferred manufacturing process usually occurs because no other process is capable of performing the task; the alternative process would require substantially longer process times or because of a reduction in scrap rates in fragile workpiece.

Using special USM tools, multitude of holes, in precise patterns can be produced simultaneously and increases productivity without compromising quality.

A novel use for ultrasonic machining has been the multistep process for fabricating silicon nitride (Si_3N_4) turbine blades. The first step in manufacturing the blades uses a shaped tool to ultrasonically trepan the blade blank from a billet of $Si_3 N_4$. A second shaped tool is then used to generate the rough airfoil shape by trepanning in a derection 90° to the initial cut and to a specific depth. After the cut is completed, a diamond slicing wheel is used to cut in from the sides to intersect with the bottom of trepan cut.

3.7 PROCESS SUMMARY

Limitations of the process

The major limitation of the process is its comparatively low metal cutting rates. The depth of the cylindrical holes is presently limited to 2.5 times the diameter of the tool. Tool wear increases the angle of the hole, while sharp corners become rounded. This implies that tool replacement is enential for producing accurate blind holes.

Due to the problem of fewer active grits coming under the tool's centre on account of in effective slurry distribution, the bottom of a cavity cannot usually be machined flat.

Example : *A drill is required to be made in 5 mm thick tungsten carbide sheet. The slurry is made of 1 part of 320 grit (15 micron radius) boron carbide mixed with* $1\frac{1}{4}$ *parts of water. The static stress is 1.4 kg/cm*^{2} *and the amplitude of tool oscillation is 0.025 mm. The machine operates at 25,000 circles/sec. The compression fracture strength of WC is 225 Kg/mm*^{2}*. Calculate the time required to perform drilling. Assume that only one pube out of 10 pulses are effective.*

Solution :

From equation, $$V = \frac{2}{3} f\, N\pi\, (2R\delta)^{3/2}$$

Metal removal rate is given by

$$V = 5.9 f (Ry_o)^{\frac{1}{2}} \left(\frac{\sigma}{H}\right) \text{ mm/sec.}$$

Here, it is given that $\sigma = 1.4 \text{ kg/cm}^2 = 1.4 \times 10^{-2} \text{ kg/mm}^2$

$$R = 15 \text{ micron} = 15 \times 10^{-3} \text{ mm}$$
$$y_o = 0.025 \text{ mm}$$
$$f = 25{,}000 \text{ cps}$$
$$\text{Rube efficiency, } \eta = 0.1$$
$$\text{Fracture strengh} = 225 \text{ kg/mm}^2.$$

Thus, $$H = \pi \times 225 \text{ kgf/mm}^2$$

Hence, $$V = 5.9 \times 0.1 \times 25000 \times \frac{1.4 \times 10^{-2}}{\pi \times 225} \times (15 \times 10^{-3} \times 25 \times 10^{-3})^{0.5}$$
$$= 57 \times 10^{-3} \text{ mm/sec.}$$

Since plate thickness is 5 mm, the time required for machining

$$= \frac{5}{5.7 \times 10^{-3}} \text{ sec.} = 14.6 \text{ min.}$$

REVIEW QUESTIONS

1. Discuss the effect of frequency and amplitude of vibration on material removal rate in USM process.
2. Define "ultrasonics" and describe the process in which these are used to machine the material.
3. What are the constituents of slurry used in ultrasonic machining ? Name the characteristics of a good suspension media. Which fluid satisfies most of these requirements ?
4. Discuss the hypothesis proposed by Shaw regarding the mode of material removal in ultrasonic machining and obtain an expression for machining rate. What are the assumptions on which this expression is based ? How far are these assumptions valid ?
5. Sketch the schematic view of tool feed system used in USM process.
6. Discuss the effects of the following parameters on the rate of material removal and surface finish obtainable in ultrasonic machining.
   (*i*) Amplitude and frequency of vibration.
   (*ii*) Abrasive grit size.
   (*iii*) Static load.
7. Explain the functions of abrasive slurry used in USM process. List the important characteristics of the abrasives used in USM. Name at least 3 common types of abrasives for use in USM.
8. Calculate the depth of indentation produced on a glass surface in ultrasonic machining by the throwing action of abrasive grain of 100 μm diameter. The following data are available :

Amplitude of vibration	0.1 mm
Frequency	20 kc/sec
Abrasive density	3.0 kg/m^3
Yield strength of glass	4.0×10^{11} N/m^2

Outline a method by which the volume rate of material removal could be computed.

❒❒❒

Ultrasonic Welding (USW)

4.0 INTRODUCTION

Ultrasonic welding is a solid state welding process wherein coalescence is produced between a wide range of similar and dissimilar combinations of metals by the local application of high frequency vibratory energy in very short cycle times and without melting the materials being joined.

Actually, the workpieces are clamped together under a modest static force normal to their interface and oscillating shear stresses of ultrasonic frequencies are applied approximately parallel to the plane of interface for about one second or even less.

Ultrasonic welding is currently used in the automotive, aerospace and electronics industries as an alternative to adhesive bonding, resistance welding, soldering and mechanical fastening.

4.1 PRINCIPLE OF OPERATION

First a power supply converts low frequency electrical energy into high-frequency electrical pulses. These are, in turn, converted to mechanical vibrations by means of a transducer.

A coupling system transmits and applies the high-frequency mechanical vibrations to the weld interface under a clamping force.

The clamping force produces a uniform, symmetrical stress that radiates into the material from the clamping point. As the lateral welding force is applied, the stress distribution in the material shifts to one side of the clamp point. This represents one-half of a vibration cycle. During the second half of the vibration cycle, the lateral force reverses direction and the stress distribution within the workpiece shifts to the opposite side of the clamping point. This rapid change in subsurface stresses occurs tens-of-thousands of times per second and results in the dispersion of surface oxides and films, the smoothing of surface irregularities through plastic deformation and ultimately, interatomic bonding through atomic diffusion.

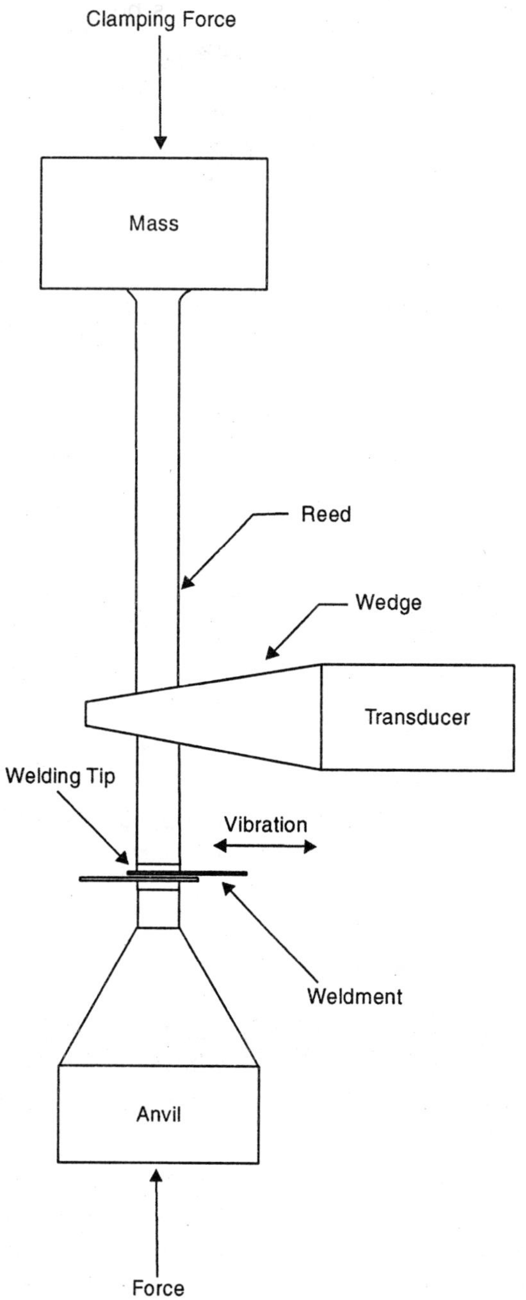

Fig. 4.1. *Diagram of the ultrasonic-welding process*

The weld interface experiences maximum temperature rise that are only 35–50% of the melting point of the material.

Because there is no melting at the weld interface, metals may be joined without the use of fluxes or shielding gases and without the chance of producing undesirable porosity or gas absorption.

Because of these capabilities, the use of USW has been expanding rapidly within the electronics and electric motor industries for wire and sheet welding. Now it is also used in aerospace and automotive companies for more critical structural applications.

4.2 WELDING EQUIPMENTS

The ultrasonic welding equipment consists of :

(*i*) Power supply

(*ii*) Transducer

(*iii*) The coupling system

(*iv*) The clamping system.

Power supply

The USW power supply is responsible for converting a 50 or 60 Hz electrical signal to a much higher-frequency signal. This is accomplished through the use of a frequency generator and amplifier. The frequency generators used for various welding applications are capable of operating at frequencies from 10 to 75 KHz.

The second stage of the power supply amplifies this high frequency signal to a usable level for the particular job. This may range from only a few watts for light duty jobs, such as wire welding, to power levels as high as 16000 W for sheet and plate welding. Modern USW power supplies are constructed with solid state circuitry, which assures reliable and stable operation over long periods.

Transducer

The role of the transducer is to convert the electrical energy from the power supply into the mechanical energy that will perform the welding. There are two types of transducers used in USW, depending upon the application. One is a magnetostrictive transducers and the other is a electrostrictive transducers (Piezoelectric transducers).

Magnetostrictive materials change length under the influence of varying magnetic flux density. In this transducer, a heavy-gauge wire is wrapped around the laminated stack of nickel sheets to form an electromagnet. Every time a pulse of current is passed through the wire, a magnetic response is induced within the stack. This results in a momentary change of about 0.005 mm in the overall length of the stack. The magnetostrictive transducer is generally used for applications requiring lower powers. It can be characterised as a low-efficiency device, although it is very rugged and long-lived.

The piezoelectric device is made of a stack of ceramic disks such as lead zirconate titanate, piezoelectric transducers respond directly to an electrical signal, that is, the stack changes length slightly as current is passed directly through it.

Piezoelectric transducers can be used for high-power applications. However because the ceramic material is brittle, it is far more fragile than the magnetostrictive transducer.

Both types of transducers require either air or water cooling to remove waste heat generated from their oscillations.

Coupling system

The coupling system transmit the high-frequency vibrations from the transducers to the workpiece. There are two methods for transmitting energy to the weld : the wedge-reed system and the lateral drive system.

The wedge-reed system, which is applicable only to spot welding, consists of a transducer that drives a wedge-shaped coupler attached to a vertical reed, as was shown in figure 4.1. When the transducer vibrates the reed, nodes and antinodes are produced along the length of the reed. The length of the reed is selected so that at a particular operating frequency, a node (point of maximum displacement) will occur at the welding tip. If an application requires very high power, several transducers can be linked to the reed at the proper nodal points.

The reed and coupler must be made of a material which is good fatique resistance. Titanium, monel or stainless steel are used for this purpose. The clamping force is provided by tip as well as anvil. The function of anvil is to hold the bottom piece of metal stationary against the lateral force of the vibrations.

Wedge reed systems are able to provide very high clamping forces but with a low amplitude, meaning that this type of system is capable of high-power applications such as large-area bonding, heavy section welding, and stranded wire welding.

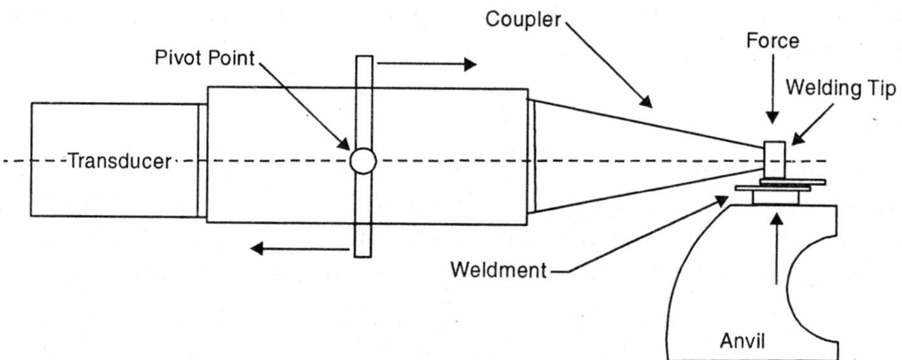

Fig. 4.2. *Lateral-drive ultrasonic welding system*

Lateral drive system consists of a horizontally mounted transducer attached to a coupling horn and tip as shown in figure 4.2. This produces a tip motion that is also parallel to the weld interface. The clamping force is generated from the pivoting motion of the ultrasonic assembly. This system can be used for either low-power spot welds or continuous seam welding.

In the lateral-drive system, energy from the transducer first travels through a tapered coupler or horn. The horn is designed with a decreasing cross-sectional area, which results in an increased amplitude at its output end. In effect the horn acts as a mechanical step-up transformer with respect to welding tip motion.

Energy from the horn is coupled into the welding tip made of materials such as tool steel or nickel alloys. The lateral-drive system is not only capable of performing spot welds, but with the proper welding tip, it can produce high speed continuous lap seam welds. This is accomplished through the use of a rotating roller tip.

Because of the low clamping force and high mechanical amplitude inherent in lateral-drive systems, this method is most often limited to low-thickness application.

Clamping system

Clamping systems provide the force being applied through either the welding tip or the workpiece itself as well as providing the static force necessary to held the workpiece together during the weld cycle. The amount of clamping force required for a particular job increases when welding softer materials or as the weld contact area increases. Simple spring system are used where power is low. Pneumatic system are used for medium power, and hydraulic systems are used for the highest powers.

4.3 PROCESS PARAMETERS

The establishment of a set of process parameters is based on the workpiece material thickness, ambient material temperature, surface condition and material properties. Therefore, if the alloy, surface finish, or weld geometry are changed, the parameters will also have to be adjusted. The parameters that are adjustable in USW are the power, clamping force, welding time and frequency.

Power

For ultrasonic spot weld, amount of power can range from a few watts for wire welding, to as much as 16000 W for heavy section welding. Seam welds are performed with power levels that range from 100 to 500 W. As power is increased, welding thickness ability is also increased and cycle time is decreased.

Clamping force

The amount of clamping force used for ultrasonic welds can vary from less than 100 gms to thousands of kilograms. Clamping forces must be increased as the contact area of the weld zone increases. For example, lighter forces would be used for wire-welding applications and larger forces would be used for plate-welding. Insufficient force can result in tip sticking; too much force needlessly increases power requirements.

Welding time

Ultrasonic spot welding is usually performed in welding cycles of 0.05 – 1 sec. each. Extended welding times are an indication that not enough power is being used. Insufficient power, and subsequently longer cycles, will increase the possibility of tip sticking, increase heating of the workpiece and increase the chances of marring the part surface at the welding tip location. Therefore it is desirable to use the highest possible power levels to minimize the welding cycle time.

The parameter of time is controlled by traverse rate in ultrasonic seam welding. The traverse rate determines the amount of energy delivered per unit length of weld and thus determines the weld penetration. Aluminium foil 0.025 mm thick can be welded at a rate of 4.5 m/min and 0.152 mm thick aluminium at 0.46 m/min.

Frequency

Frequency is adjusted to match the natural resonance frequency of the coupling system and is then held constant to within 1–2% of its preset value via a close-loop control system. A high operating frequency results in a lower amplitude and a corresponding decrease in process efficiency.

4.4 PROCESS CAPABILITIES

Figure 4.3 illustrates the combinations of metals that are applicable to USW. Blank areas on the chart indicate combinations that either have not been successfully welded or that have not yet been tried.

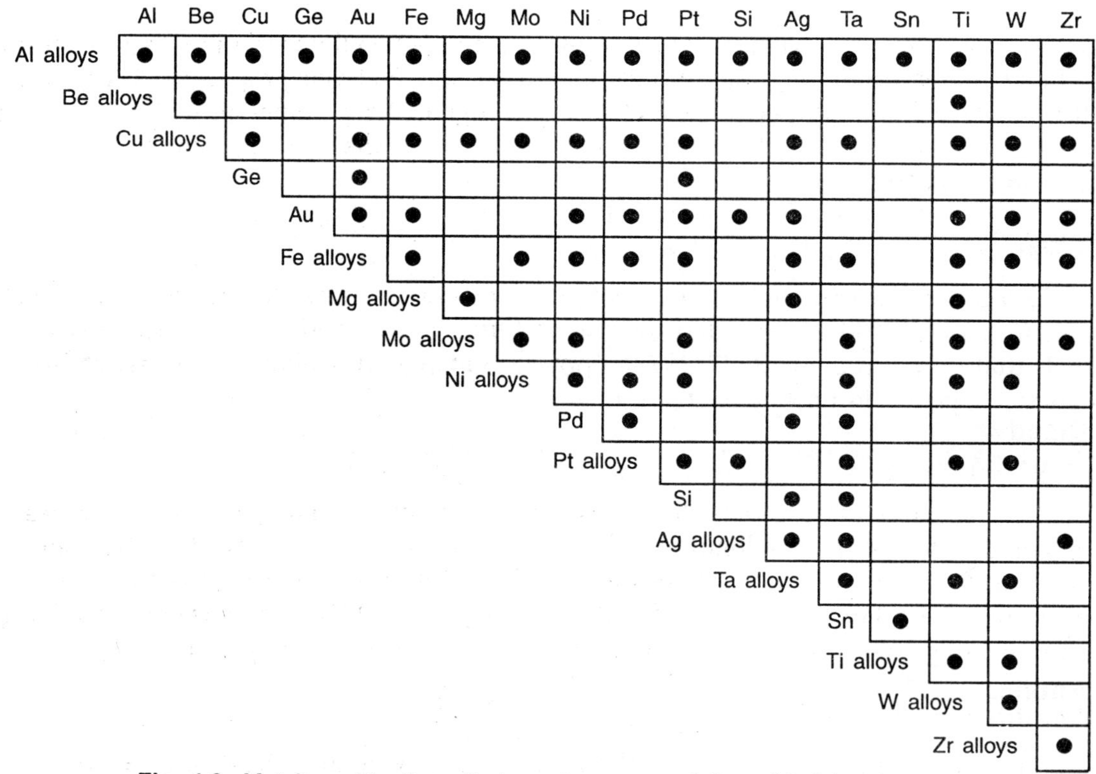

	Al	Be	Cu	Ge	Au	Fe	Mg	Mo	Ni	Pd	Pt	Si	Ag	Ta	Sn	Ti	W	Zr
Al alloys	●	●	●	●	●	●	●	●	●	●	●	●	●	●	●	●	●	●
Be alloys		●	●			●										●		
Cu alloys			●		●	●	●	●	●	●	●		●	●		●	●	●
Ge					●						●							
Au					●	●			●	●	●	●	●			●	●	●
Fe alloys						●		●	●	●	●		●	●		●	●	●
Mg alloys							●						●			●		
Mo alloys								●	●		●			●		●	●	●
Ni alloys									●	●	●			●		●	●	
Pd										●			●	●				
Pt alloys											●	●		●		●	●	
Si													●	●				
Ag alloys													●	●				●
Ta alloys														●		●	●	
Sn															●			
Ti alloys																●	●	
W alloys																	●	
Zr alloys																		●

Fig. 4.3. *Metal combinations that can be successfully welded by ultrasonics*

The materials to weld by USW are aluminium, copper, gold, silver, platinum, thier alloys and steels. Refractory and some high-strength metals such as molybdenum, tungsten, zirconium, tantalum, columbium, titanium and nickel present some difficulties in that thicknesses are limited to thin foils and welding tip life is reduced.

The welding thickness is limited by the thickness of the top-joint member. Multiple layers of foil can be spot or seam welded together depending upon the thickness of the layers to be welded and the power of the machine. As many as 20 layers of foil have been successfully welded in production application.

Aluminium sheet can be welded in thicknesses upto 2.54 mm and refractories are limited to maximum thicknesses of 0.38 – 1.02 mm. Wires from 0.025 to 0.76 mm in diameter are within the capabilities of USW while the minimum thickness of sheet metal that can be welded reliably is 0.005 mm.

The ultrasonic welds exhibit greater ductility in comparison to resistance welding because no melting has occurred. Furthermore, ultrasonic welding offers cost and quality advantages because of simplified surface preparation requirements, lower energy consumption and the ability to successfully weld workpieces that are grossly different in composition, properties or thickness.

4.5 PROCESS SUMMARY

Advantages

(*i*) Surface preparation is not critical.
(*ii*) No defects are produced from arc, gases and filler metals.
(*iii*) Dissimilar metals having vastly different melting points can be joined.
(*iv*) Minimum surface deformation results.
(*v*) Very thin materials can be welded.
(*vi*) Thin and thick sections can be joined together.
(*vii*) Due to low temperature involvement, the characteristics of the materials are not altered and are continued through the weld zone.
(*viii*) To weld glass is impossible by any other means.

Disadvantages

(*i*) This process is restricted to joining thin materials—sheet foil and wire. The maximum is about 3.0 mm for aluminium and 1 mm for the harder metals.
(*ii*) Materials being welded tend to weld to the tip and anvil.
(*iii*) Due to fatique loading, the life of equipment is short.
(*iv*) Hard materials will fatique under the stresses necessary for welding.

Applications

The most common application for USW traditionally lie within the electronic, electrical and defence industries. Applications vary from welding wires and containers for microcircuits to attaching wires and contacts to motors and switches. It is used for hermetic sealing of volatile substances and in-fabricating nuclear fuel elements. Continuous seam welding to assemble components of corrugated heat exchangers is done by USW.

REVIEW QUESTIONS

1. What is the difference between ultrasonic machining and ultrasonic welding process ?
2. What is the mechanism involved in the ultrasonic welding ?
3. What role is played by the transducer in USW and how significant it is ?
4. How welding time is affected by clamping force and welding area ?
5. What is the potential application of USW ?

❑❑❑

Water Jet Machining

5.0 PROCESS PRINCIPLES

Water jet machining is a process used primarily to cut and slit porous nonmetals such as wood, paper, leather and foam.

Water jet machining removes material through the erosion effects of a high velocity, small-diameter jet of water. In early 1900s in a steam plant, workers observed that a pinhole leak developed in a high-pressure steam line resulting a steam jet had sufficient power to quickly and clearly cut through objects such as a wooden broomstick. Later on Norman Franz patented the technique for producing a coherent, high-velocity stream of water. This principle was introduced in the industry as a new cutting tool in the early 1970s.

In fact if a jet of water is directed at a target in such a way that, on striking the surface the high velocity flow is virtually stopped, then most of the kinetic energy of the water is converted into pressure energy. In fact, in the first few milliseconds after the initial impact of the jet on the target, before the lateral flow of water is initiated, a transient pressure as much as three times the normal stagnation pressure, may be generated. Erosion occurs if the local fluid pressures exceed the strength of the bond binding together the materials making up the target. In other words, liquid jet cutting removes material primarily by the mechanical action of a high velocity stream impinging on a small area, whereby its pressure exceeds the flow pressure of the material being cut.

A number of studies using liquid jets have been made for cutting coal or rock. Farmer and Attewell reported the results of water jet impinging on sandstone. The system used was a pulsed jet with velocity upto 500 m/s, and the effects of velocity on penetration were reported. The effect of varying stand-off distances at pressures upto 92 MN/m^2 were reported for jets with and without a polymer additive. Pulsed water jets have been used in rock excavation and machining aluminium and lead. Cutting capability at pressures upto 10000 atm. was reported for a wide variety of target materials, including wood, lead, rubber, aluminium, copper and steel. A more recent study has been reported on the effectiveness.

5.1 JET CUTTING EQUIPMENT

The figure 5.1, shows a schematic diagram of water jet cutting system. The pumping unit (oil pump) is driven by an electric motor. The oil drawn from a reservoir, is pumped

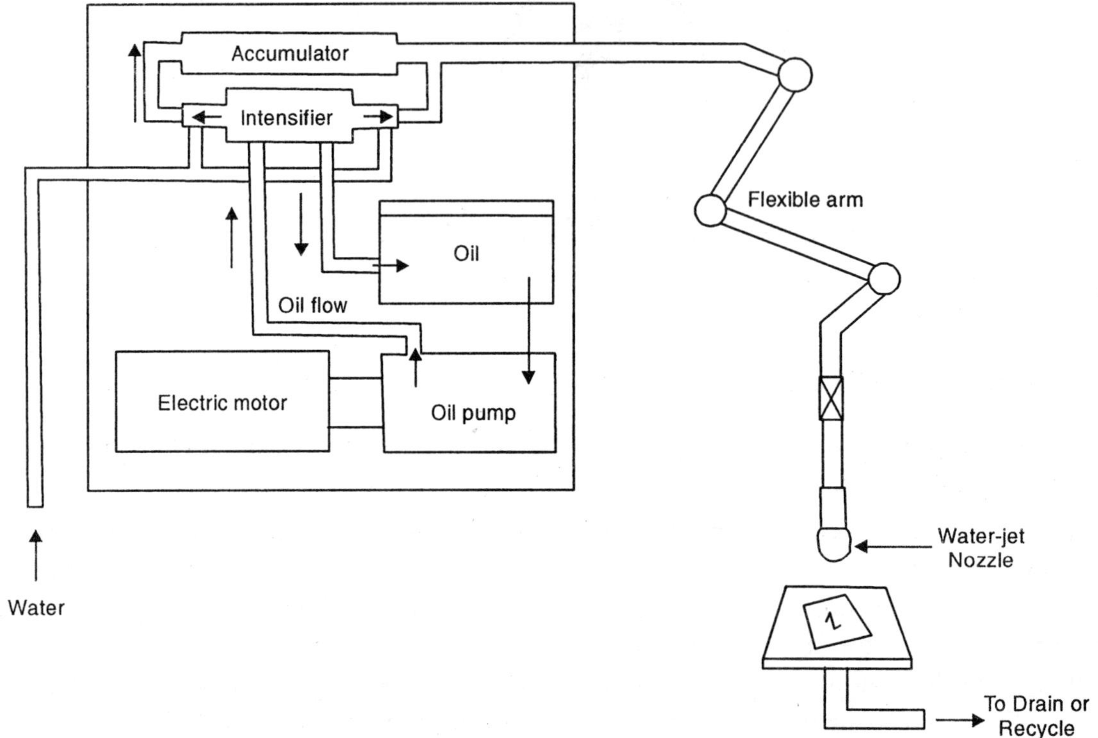

Fig. 5.1. *Schematic diagram of water jet machining system*

to an intensifier which uses low pressure oil to produce very high pressure water. The intensifier that acts as a high pressure pump, produces water pressure as high as 40 times that of the oil. Water pressure (P_w) can be determined from the following equation :

$$P_w = (P_0 \times A_0)/A_w$$

where P_0 is oil pressure, A_0 is oil piston area and A_w is water piston area.

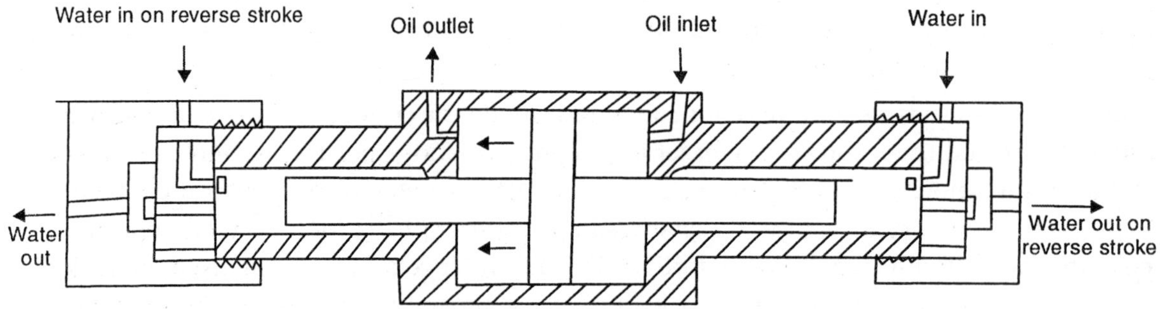

Fig. 5.2. *High pressure oil intensifier*

Adjustment of the high pressure water is easily accomplished by regulating the low-pressure oil. Pressure regulation of the oil is accomplished with readily available hydraulic components.

To minimize pulsation in water flow, a high pressure accumulator (*i.e.*, a pressure vessel to store high-pressure water to give smooth outflow) is used. This action maintains output pressure variations of not more than ± 5%. High pressure water is transported to nozzle through the rigid high pressure tubing and rigid connectors.

A valve is installed in the high pressure water line to provide on-off control of the water flow. Because pressure-compensated hydraulic pumps are used in WJM systems, turning off the water flow will not cause an increase in line pressure. On-off valves can be installed at any point along the high-pressure water line and may be operated either manually or electronically.

Nozzles

Nozzles meant to convert the high pressure liquid to a high velocity jet. For minimum erosion, the nozzle material should be extremely hard. Yet, to allow the formation of a smooth contour, the material should be ductile and easily machinable. Proper design of nozzles has made it possible to discontinue the use of long chain polymers to lower down friction in fluid flow. The nozzles are made of synthetic sapphire which is wear resistant but easily machinable. Presence of foreign particles (say dirt) in water results in failure of nozzle by chipping. Sometimes, constriction of nozzle by mineral deposits also results in nozzle failure. Life of a sapphire nozzle is usually 250–500 hrs. Internal diameter of nozzle usually ranges from 0.07 to 0.50 mm.

Catchers

Water jet outside the nozzle, travels at a very high speed (usually more than sound velocity). To minimize the exposed length of a jet, from safety point of view and also to minimize the process noise, a catcher (a slot type or tube type) is used.

The tube catcher, the most common type, consists of a 300 to 600 mm long tube that is attached to a draining hose. The length of the tube is sufficient to allow the water jet to break up completely before it reaches the bottom of the tube. When it would be awkward to manipulate a catcher of this length behind a part, as with the hand-held device, a short tube catcher can be used. This uses a hard, replaceable impingement insert to break up the jet quickly. For slitting system incorporating multiple or adjustable nozzles, a slot type catcher is used. Although the slot-type catcher may be the most convenient in this situation, it is less effective in reducing noise levels.

5.2 PROCESS PARAMETERS

The process variables which controls the WJM process are pressure, nozzle diameter, traverse rate and the stand-off distance. Of these, stand-off distance is the least critical.

Cutting is usually performed with a nozzle to workpiece stand-off distance of less than 25 mm and optimally 3 mm. Because there is little change in the shape or diameter of the jet within 25 mm of the nozzle, the stand-off distance can be considered a fixed parameter. Hence only three parameters-pressure, nozzle diameter and traverse rate are varied, depending upon the material and the thickness being cut. In general, as the power of the water jet is increased, the ability to cut faster or to cut thicker materials is also increased. This is accomplished by increasing the pressure, increase the nozzle diameter, or decreasing the traverse rate.

The relationship between nozzle diameter, pressure, flow rate and pump size is shown in figure 5.3. This chart assumes a 1.0 nozzle orifice co-efficient. The most accurate value for this coefficient for WJM applications is 0.7.

5.3 PROCESS CAPABILITIES

Cutting and slitting

A wide range of materials can be cut with WJM, but because erosion is the method of removal, materials must be porous, fibrous, granular, or soft.

In this process cutting rate can be extremely high and hence thickness may be built up by multilayer cutting, thus slowing the cutting, rate, or the cutting nozzle may be moved by a cam-driven actuator.

For shape cutting with WJM, predrilling is not required because the jet can pierce material in any location and then proceed to cut in any direction. This omnidirectional cutting ability allows complex shapes with almost any radius to be easily fabricated.

The thick material can be cut with multiple passes without degrading the cut quality. If the first pass can produce a well-defined slot in the material, then deeper cut takes place in second pass. Less energy is used in multipass cutting than a single pass with more powerful jet.

The surface finish of cut edges is good, with no burrs and no heat damage. However, surface finish and edge straightness degrade as material becomes very thick or when cutting speeds are too fast.

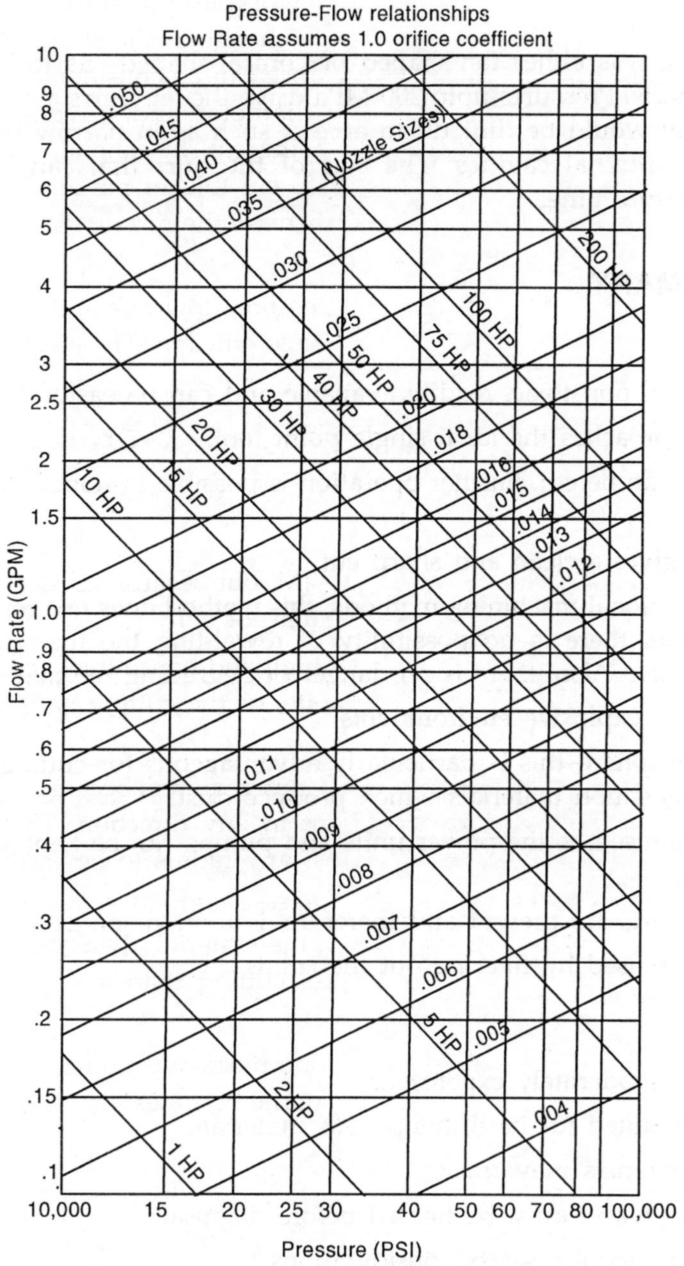

Fig. 5.3. *The WJM pressure/flow relationships*

Cable stripping

Water jet cannot cut readily the metal and this principle is used in cable stripping. Thus during cable stripping, jet easily pierces and cuts insulation without harming the underlying metallic cable. Cable stripping is performed with water pressures of 69–200 MPa, much lower than WJM cutting and insulation thickness as great as 12.7 mm can be cut.

Deburring

Water jet deburning was either fan-shaped or conical-shaped jets to break or erode burrs from nonferrous parts. Pressures upto 200 MPa drive the jet. Burrs are easily removed from areas that normally would be difficult to access, such as in narrow openings, blind holes and line-of sight internal cavities. The size of the burr that can be removed is proportional to the exposure time.

5.4 PROCESS SUMMARY

Advantages

(*i*) Water is cheap, non toxic, readily available and can be easily disposed.

(*ii*) Water jets approaches the ideal single point tool.

(*iii*) Any contour can be cut. Further operation is possible in horizontal and vertical planes.

(*iv*) The process gives a clean and sharp cut.

(*v*) Unlike conventional machining methods, this method does not generate heat. As a consequence, there is no possibility of rewelding the material behind the advancing cutter. Also, there is no danger of degrading the material thermally.

(*vi*) Best suited for explosive environments.

(*vii*) Dustless atmosphere–this is particularly advantageous for cutting asbestos and glass fibre insulation materials which produce dust.

(*viii*) Noise is minimised as the power units and pumps can be kept away from the cutting point.

(*ix*) No moving parts are present and therefore, less maintenance is required.

(*x*) Fluid can be reused by filtering out the solids.

Disadvantages

(*i*) Equipment is moderately expensive.

(*ii*) It is not well suited for hard, nonporous materials.

(*iii*) The brittle materials may crack.

(*iv*) Contaminated water must be treated before disposal.

(*v*) High pressures require safety considerations.

REVIEW QUESTIONS

1. What is the working principle of water jet machining ?
2. Write a note on the special features of the equipment used in the method of machining.
3. Give the practical applications of water jet machining.

❑❑❑

6

Chemical Machining

6.0 INTRODUCTION

The practice of metal removal using chemical solutions has a history dating back approximately 4500 years. In 2500 BC Egyptians used citric acid solutions to decoratively etch copper jewellery. 1000 years ago in State of Arizona (USA), fermented cactus juice were used to etch jewellery.

Chemical machining remained primarily an artist's tool until the late 1930's when American industry began a limited use of chemicals in manufacturing. In the 1940s, North American Aviation, Inc. patented a process called Chem. Mill. Which was used for the fabrication of aircraft using panels. Also in the 1940's chemical machining began to emerge as the preferred technique for selective removal of copper to produce printed circuit boards. The practice of 'blanking' thin metal parts through the use of chemicals was begun in the early 1950's but did not come into widespread use until the early 1960's.

6.1 FUNDAMENTAL PRINCIPLES

In chemical machining process, the metal is removed by spraying the etchant solution on the machining zone of the workpiece or by dipping the workpiece in the etchant solution. The metal is gradually transformed into metallic salt by chemical reaction and is ultimately removed. The metal removal rate is controlled by controlling the concentration, composition and operating conditions of the etchant solution. In selective metal removal process, the unmachined portions of the workpiece are protected by an etchant resistant material, known as "maskant" or "resist".

The workpiece is first coated with the maskant material which may be cut and peel, photoresist or screen-print type. Depending on the type of maskant used, the surfaces to be etched are then located by scribing the maskant or by using photographic technique or by screen-print technique.

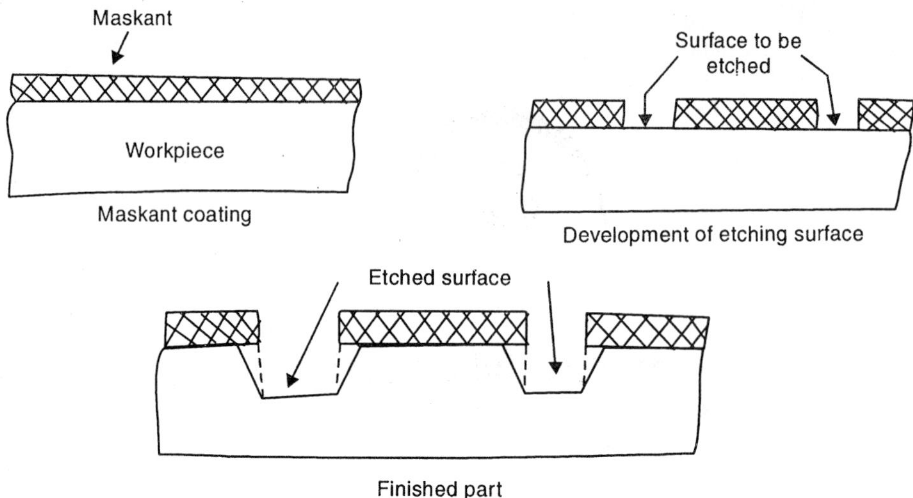

Fig. 6.1

The chemical-machining process can be classified as two major CHM processes such as chemical milling and chemical blanking depending upon the intended application and the method of applying the chemical resist material.

Chemical milling is defined as the process of chemically eroding material to produce 'blind' details (pockets, channels etc.) or to remove material from all surfaces of a part for the purpose of weight reduction.

Chemical blanking is a process for producing details that penetrate the material entirely (holes, slots, etc.), or the process of blanking complete parts from a sheet of material by chemically etching the periphery of the desired shape.

The chemical machining process have the following common processing steps :

(*i*) Preparing : precleaning.

(*ii*) Masking : application of chemically resistant material (if selective etching is desired).

(*iii*) Etch : dip or spray exposure to the etchant.

(*iv*) Remove mask : strip remaining mask and clean.

(*v*) Finish : inspection and post processing.

Precleaning operations are necessary with chemical machining to assure the proper adhesion of the masking material to the workpiece. Proper precleaning lowers the possibility of maskant debonding which, in turn, can result in stray etching. Depending upon the type of mask material, the required depth of cut, and the workpiece material, the cleaning techniques may vary from a simple solvent wipe to more thorough processes such as flash etching, vapour degreasing, or alkaline etching.

For parts requiring selective etching, a chemically resistant mask is applied to the part and inspected for incomplete coverage and lack of adhesion. After masking, the parts are exposed to a chemical etchant that uniformly removes material from all exposed surfaces.

To ensure uniform material removal, the etchant is either continuously sprayed on to the part or the part is submerged in a tank of agitated etchant. These methods prevent uneven etching by constantly circulating fresh etchant over the part's surfaces. However, too much agitation of the etchant can create areas of cavitation or stagnation that will result in ridges, waviness or grooves in the etched surface.

During the etching portion of the process, the material erosion progresses both inward and laterally from the exposed surface. The distance etched into or through the part is known as the depth of cut, and the distance etched laterally under the mask is referred to as the undercut.

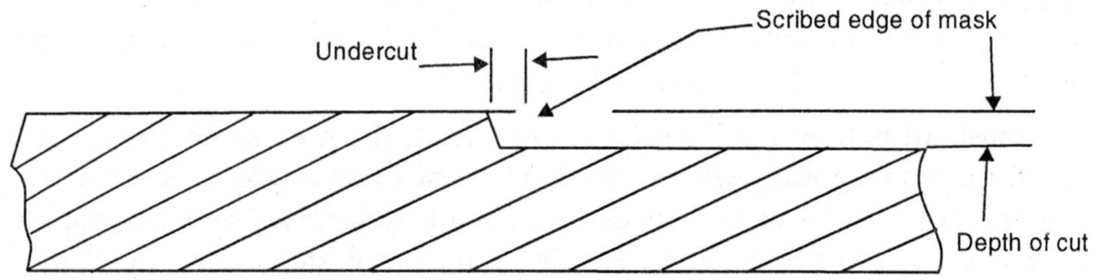

Fig. 6.2. *Chemical-machining undercut*

The amount of undercut that occurs in a particular application is a function of many factors including the depth of cut, the type and strength of the etchant and the workpiece material. The compensation for undercut must be made during selective masking to ensure the proper final size and is given by

$$\text{Etch factor} = \frac{\text{Undercut}}{\text{Depth of cut}}$$

The actual depth of the etch can be precisely controlled by simply controlling the immersion time in the etchant. It must also be recognised that during the etching process, gases are released as bubbles. If they become trapped, these bubbles can insulate the workpiece from the etchant and consequently result in non-uniform material removal. Therefore, to avoid the problem, workpieces are usually tilted or oriented to encourage the bubbles to escape.

After etching, the mask is removed by mechanical and/or chemical stripping. The mechanical techniques may be necessary for the thicker, more durable masks, while the chemical techniques are generally adequate for the thinner masks, or for thin, delicate parts that could be easily damaged.

The final CHM process step is either inspection or on to further processing by other techniques until finally assembled into a product.

6.2 PROCESS PARAMETERS

The chemical machining process is usually governed by the composition and temperature of the etchant, the method of circulating the etchant, the type of maskant and the method of applying the maskant.

Etchant resistant material (Maskants)

On the basis of the process requirements, the etchant resistant materials which are commonly used in practice can be classified into the following groups :

Cut and peel resists

These are first applied to the entire part by either dip, spray or flow coating. These coatings are applied in thicknesses that range from 0.025 to 0.13 mm. Because the coatings are so thick, they can withstand exposure to the etchant for extended periods so that workpieces can be etched as deep as 13 mm.

The maskant is then selectively removed from the area to be etched by hand scribing using a template as a guide. After scribing, the unwanted maskant is manually peeled away, exposing the areas to be etched. Because of manual scribing and the maskant is so thick image accuracy is never better than ± 0.13 – ± 0.75 mm, depending upon the size and type of feature being produced.

This mask is the only mask that can be easily described to produce step etching.

The cut and peel technique is most suitable for the parts that are large, require etch depths deeper than 1.5 mm per side, low quantity, require multistep depths, or are contoured on their surfaces. The maskants commonly available are vinyl, neoprene and butyl base material.

Photoresist masks

Photoresist masking is commonly known as photochemical machining (PCM) and is used to produce intricate and precise shapes by using a photographic image to selectively cure the photoresist mask on a workpiece.

The first steps for PCM involve generating the artwork that will be used to create the desired image in the workpiece. The artwork is either handdrawn or computer generated 2 to 20 times life size. Compensation values are used in the design to account for the effects of undercut.

After the design is complete, the artwork is photographed and reduced to produce a highly accurate, photographic master transparency (photometer). If the application requires that many identical parts are to be blanked out of one sheet of material, a precision step and repeat camera will be used to provide many duplicate images on a single photomaster. An alternative to producing a photomaster from drawing is to use a laser pattern generator.

After the photomaster is created, the part must be prepared by a thorough cleaning to remove all dirt and oxides. After cleaning, the part is coated with a thin layer of a light-activated, etchant-resistant material (photoresist) and baked dry. Because the most commonly used photoresists are cured by exposure to ultraviolet light, the photoresist can be accidentally cured by exposure to sunlight or standard fluorescent lights.

After the workpiece has been prepared, a strong ultraviolet light source is used to selectively expose the photoresist while the photomaster and workpiece are held together in intimate contact. After exposure, the photoresist is developed to remove the coating from all areas where etching is desired.

After the photoresist has been developed, the parts are chemically etched, stripped of residual photoresist, and inspected.

To prevent small parts from prematurely dropping out of the sheet from which they are being blanked, small attachment tabs may be incorporated into the design to remain after etching to secure the parts for easy handling.

These resists are used for

(*i*) Thin materials upto 0.8 mm thick.

(*ii*) Parts requiring dimensional tolerances of the etchant resistant image tighter than ± 0.1 mm and upto 1 × 1.5 m section.

(*iii*) Automatic processing of high volume components. However, the applicability of these types of maskants is limited by their ability to etch depth upto a thickness of 2 mm.

Screen-print maskants

In this technique, a fine mesh silk or stainless steel screen, which has areas blocked-off to allow selective passage of the maskant is used. The blocked pattern corresponds to the image that is to be etched. The screen is pressed against the surface of the workpiece and the maskant is rolled on. When the screen is removed, the maskant remain on the part in the desired pattern. The maskant is ready for etching after it has been dried by baking.

Screen printing is a fast, economical masking method used for high-volume production when high accuracy is not required.

Screen printing is generally limited to :

(*i*) Parts not larger than 1 × 1 m.

(*ii*) Parts having only flat surfaces or simple moderate contours.

(*iii*) Parts where depth of etch does not exceed 1.5 mm in depth from one side.

(*iv*) Parts which do not require etchant resistant image accuracy greater than ± 0.2 mm.

Selection of maskants

The factors which greatly influences the selection of maskants are :

(*i*) chemical resistance offered by the maskants to the etchants,

(*ii*) rate of production of parts,

(*iii*) ease of removal of maskants at the end of operation,

(*iv*) shape and size of parts, and

(*v*) economics involved in the process.

Etchants

The function of etchants is to dissolve and thus, remove the metal from the working zone by converting it into metallic salts. The factors which affect the selection of an etchant for a given component are :

(*i*) Type of material to be etched.

(*ii*) Type of maskants or resists employed.

(*iii*) The metal removal rate desired.

(*iv*) The operating conditions.

(*v*) The requirement of surface finish.

(*vi*) The economics involved in material removal.

Table 6.1. Etchant Characteristics and Applications

Etchant	Concentration	Temp. (K)	Etch rate (cm/min) fresh solution	Etch factor	Metals that etchant will attack
$FeCl_3$	12 to 18 Be*	320°	0.002	1.5 : 1 to 2 : 1	Aluminium alloys
HCl,					
HNO_3H_2O	10 : 1 : 9	320°	0.002 to 0.004	2 : 1	
$FeCl_3$	42° Be*	320°	0.002	2 : 1	Cold rolled
HNO_3	10 to 15% (Vol.)	320°	0.002	1.5 to 2 : 1	steels
$(NH_4)_2S_2O_8$	0.22 gm/cm^3	300 – 320°	0.002	2 to 3 : 1	Copper and its alloys
HNO_3	12 to 15% (Vol.)	300 – 320°	0.002 to 0.004	—	Magnesium
$FeCl_3$	42° Be*	320°	0.001 to 0.002	1 : 1 to 3 : 1	Nickel
$FeCl_3$	42° Be*	325°	0.002	1.5 to 2 : 1	Stainless steel, tin
HNO_3	10 to 15% (Vol.)	320° to 325°	0.002	—	Zinc

* Baume specific gravity scale

The chemical machining process will be useful in case of chemical milling, blanking and engraving operations.

6.3 PROCESS SUMMARY

Advantages

The advantages of this process are :

(*i*) Several parts can be machined simultaneously.

(*ii*) Machining is done without any production of burrs.

(*iii*) Very thin sheets of metal can be processed with ease without distortion.

(*iv*) The machining accuracy actually increases with the decrease in metal thickness.

(*v*) The hardness of the metal to be machined has no significant effect on the process. The temperability, stress and other physical properties of the metal are not altered.

(*vi*) Materials which are brittle in nature can be processed with ease.

(*vii*) The flexibility in the design of the process is unlimited.

Limitations

(*i*) The operator skill required is relatively high.

(*ii*) The etchant vapours are very corrosive in nature. Hence, the process equipments must be kept safe from the etching and other operating equipments.

(*iii*) The maximum working metal thickness is small.

(*iv*) The parts finished by chemical machining have undercutting which is further increased with the increase in metal removal thickness.

(*iv*) Productivity is relatively low in this process. Optimum productivity is determined by the part size, metal thickness to be machined and quantity of parts required.

Applications

Chemical machining process is applied in cases where the depth of material removal is critical to a few microns and the tolerances are close to be maintained. The surface finish obtained in the process is of the order of 0.5–2 microns. This process is used to :

(*i*) Remove metal from a portion or the entire surface of formed or irregularly shaped parts, such as forgings, castings, extrusions, or formed wrought stock.

(*ii*) Reduce web thickness below practical machining, forging, casting or forming limits.

(*iii*) Taper sheets and pre-formed shapes.

(*iv*) Produce stepped webs.

(*v*) Engraving on any metal piece.

REVIEW QUESTIONS

1. What are the factors on which the selection of a resist for use in chemical machining depend ?
2. Distinguish between cut and peel resists and photographic resists.
3. What are the specific advantages of using chemical machining over electrochemical machining ? Give some practical applications of the chemical machining process.

❑❑❑

Electro-Chemical Machining (ECM)

7.0 INTRODUCTION

Electrochemical machining process is one of the latest and potentially the most useful of the non-traditional machining processes. This process relies on the principle of electrolysis for material removal. A depleting action between a conductive workpiece and a shaped tool produces a predictable erosion of the workpiece. The metal removal rate is governed by the well-known Michael Faraday's laws of electrolysis (1791–1867). The metal is mainly removed in the form of sludges and precipitates by electro-chemical and chemical reactions occurring in the electrolytic cell. The process has immense applicability in the field of forceless cutting of conventional type of material, especially when the part shapes are complicated or when several conventional machining operations are to be finished by a single stage operation of the ECM tool. Electrochemical machining is most often applied when shaped cavities are machined into alloys that are difficult to shape by conventional methods. These cavities are quickly produced by ECM by simply feeding the tool into the workpiece until the required depth is reached. Although this process has the ability to produce complex shapes in one pass of the tool, its use is often limited to mass production because of high tooling and set-up costs.

7.1 CLASSIFICATION OF ECM PROCESSES

Depending on the mechanism of anodic dissolution and the removal of the reaction products from the machining zone, ECM methods may be classified into various groups as shown in figure 7.1. The different fields of application of the ECM process are : Forming, Drilling, Cavity Sinking, Milling, Slitting, Internal Forming, Turning, Machining Internal Helical Splines, Finishing Cavities after EDM etc.

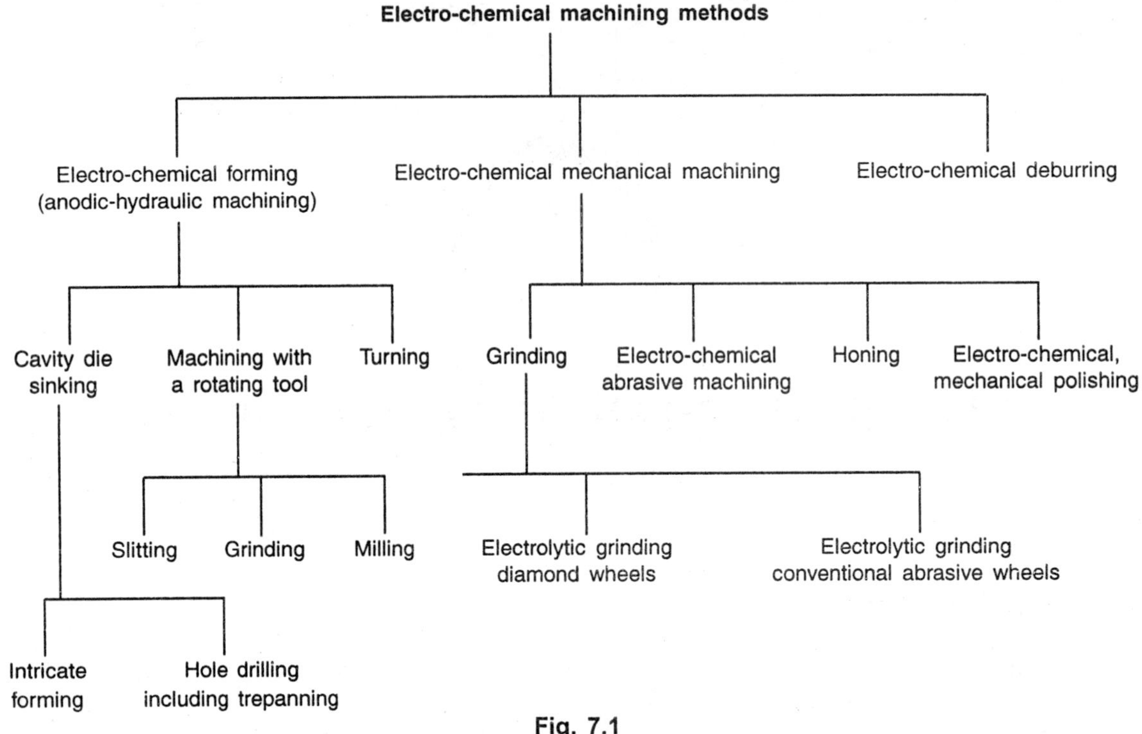

Fig. 7.1

7.2 FUNDAMENTAL PRINCIPLES OF ECM

Electro-chemical machining (ECM) is an inherently versatile process of machining because of its capability of stress free machining of various kinds of metals and alloys. It can produce shapes and cavities which are costly and extremely difficult to machine with the conventional machining processes and a true replica of the tool (or cathode) can be made on the workpiece (or anode) by the controlled dissolution of the anode of an electrolytic cell. The basic scheme of the ECM process has been shown in figure 7.2 in which the tool serving as the cathode and the workpiece serving as the anode of an electrolytic cell have been shown : An electrolyte (usually a neutral salt solution such as sodium chloride, sodium nitrate, sodium chlorate) is passed through a very small gap (0.05 to 0.3 mm) created between the workpiece (or anode) and the tool (or cathode) whereas a direct current flow is made between them.

When sufficient electrical energy (about 6 eV) is available, a metallic ion may be pulled out of the workpiece surface. The positive metallic ions will react with the negative ions present in the electrolytic solution forming metallic hydroxides and other compounds. Hence, the metal will be anodically dissoluted with the formation of sludges and precipitates.

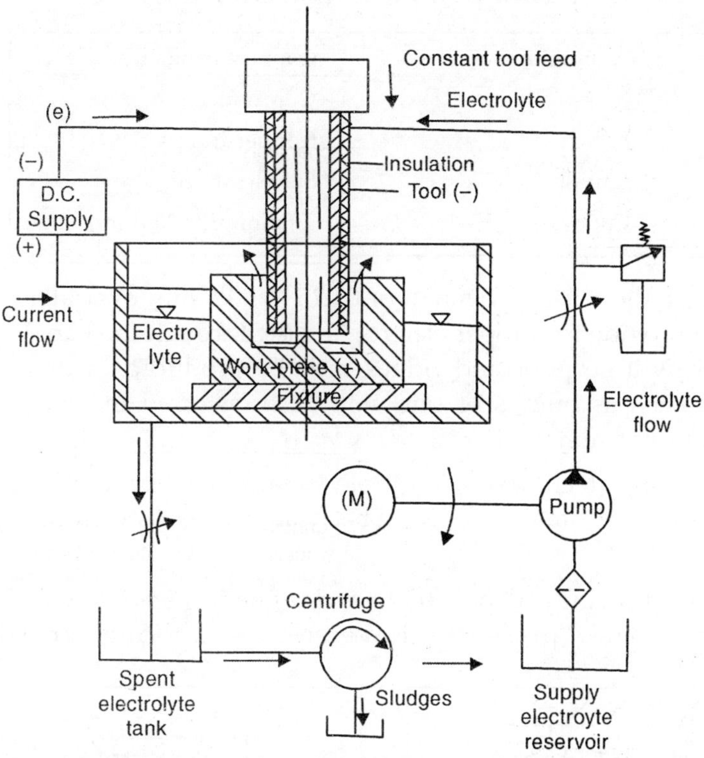

Fig. 7.2. *Basic scheme of electro-chemical machining process*

7.3 ELEMENTS OF ECM PROCESS

The elements of the process are :

(*i*) Cathode tool (of a shape which almost the mirror image of the cavity to be machined into the workpiece).

(*ii*) Anode workpiece (and means to hold and locate it near the tool).

(*iii*) Source of D.C. power (off sufficient capacity so that high current densities can be maintained between the tool and workpiece).

(*iv*) Electrolyte, a conductive liquid (and means to supply it into the gap between the tool and workpiece).

(*v*) Tool-work gap.

Cathode tool

Electrochemical machining tool material should be good electrical and thermal conductor, be easily machinable, resist chemicals, exhibit good stiffness (to resist the high electrolyte pressures) and be easily obtained. The materials that find wide applications in the manufacture of tools for ECM are given in the table 7.1.

TABLE 7.1. ECM Tool Materials

Aluminium	Copper-manganese
Brass	Cupro-nickel
Bronze	Copper-tungsten
Carbon	Stainless steel
Copper	Titanium (99% pure)

Tool design must allow for an overcut of 0.025 to 0.76 mm over all active surfaces. Tool must be designed to provide for proper electrolyte flow through the gap. If area of stagnation occur, surface finish will be poor and ridges may be produced in the workpiece surface. Defects in the tool, such as bumps or pits, will be transferred in reverse image on to the workpiece if they are not removed.

Using proper design techniques, multiple details may be machined simultaneously with one tool. As many as 50 small holes have been simultaneously produced with a multiple-tool design.

Figure 7.3 shows drawings of tools that are designed to operate with reverse electrolyte flow and variable back pressure. Reverse-flow tooling greatly reduces the occurrence of non-uniform material removal and sludge-induced short circuits. The high electrolyte back pressure that results from the reverse flow causes the hydrogen bubbles in the gap to decrease in size, thus increasing the conductivity and the feed rate.

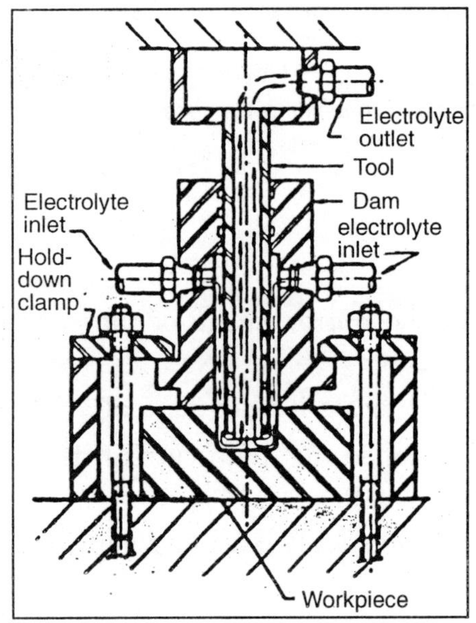

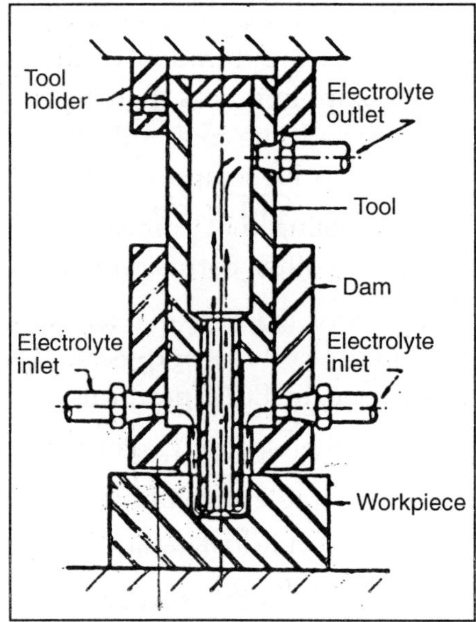

Fig. 7.3. *Two examples of reverse-flow tool designs used to control back-pressure and accelerate ECM removal rates*

Anode workpiece

There is no restriction on the nature of the work material except that it must be a good conductor of electricity. The chemical characteristics of the workmaterial, however, do affect the material removal rate. The removal rate is proportional to the atomic weight and inverse of the valency of work material.

D.C. Power and control system

The process needs low voltages of the order of 2 to 20V and in rare cases upto 30V. Normal current requirements are as high as 800 amp/cm^2 of the workpiece area to be machined. Now a days even a higher currents are being considered such as 40000 amp. A computer, Controlled, 20000 amp tri-ram ECM system is used for making turbine blade.

The tool-to-workpiece gap (0.25 mm approx.) must be maintained constant throughout the ECM cycle. This is accomplished by utilizing a servo-drive on the tool feed axis.

Occasionally, deposits may build up in the gap or the tool may come too close to the workpiece resulting in a short circuit. Because of the high current used, a short circuit will usually destroy both the tool and the workpiece. The amount of damage sustained by a short circuit is a function of the energy of the spark that is produced, *i.e.*, it is determined by the current, voltage and spark duration. The only parameter that can be controlled for short circuit protection is the duration of the spark. A solid-state devices are used which disconnect the power supply in less than 10–20 μ sec. after serving a short circuit. If power is removed in short period, the tool and workpiece will receive a spark so small that they will probably not have to be scrapped.

The machine or tooling that comes in contact with the electrolyte must be made of stainless steel, plastic or other corrosion-resistant materials.

Electrolyte

The electrolyte used in ECM performs many functions, such as :

(*i*) Completing the electric circuit between the tool and the workpiece.

(*ii*) Allowing desirable machining reactions to occur.

(*iii*) Carrying away heat generated during the chemical reactions.

(*iv*) Carrying away products of reaction from the zone of machining.

An effective and efficient electrolyte should have the following characteristics :

(*a*) High electrical conductivity.

(*b*) Low viscosity and high specific heat.

(*c*) Chemical stability.

(*d*) Resistance to formation of passivating film on work surface.

(*e*) Non-corrosive and non-toxic in nature.

(*f*) Inexpensive and readily available.

Electrolyte can be divided into two classes–those composed of inorganic salts that produce insoluble by-products (known as sludging electrolytes) and those composed of acids or alkalis that result in by-products that do go into solution (known as nonsludging electrolytes).

Sodium chloride, potassium chloride, sodium nitrate, and sodium chlorate are some of the inorganic salts used as sludging electrolytes. The selection of electrolyte is based upon the workpiece material, the tool material and the application. Sodium chloride is the most commonly used electrolyte because of low cost and because its conductivity is stable over a broad pH range. However, sodium chloride is also very corrosive, produces a large amount of sludge, and cannot be used on tungsten carbide or molybdenum.

Sodium nitrate is also popular. This material is much less corrosive and is good for the machining of aluminium or copper. Sodium nitrate however has the problem of causing passivity of the workpiece (an oxide layer buildup which slows the electrolytic process) and does not produce surface finishes as fine as sodium chloride.

While nonsludging electrolytes such as sulphuric acid or sodium hydroxide do not produce the large amounts of sludge that the inorganic salts do, their chemistry must be closely controlled to realize repeatable results, otherwise they have a tendency to plate the tools. For certain specific materials (such as molybdenum) the nonsludging electrolytes produce the best results.

All electrolytes must be cooled to remove waste heat and maintain temperature. The electrical conductivity of an electrolyte changes as a function of temperature. *e.g.*, sodium chloride increases its conductivity by 100% when the temperature is increased from 24 to 71°C.

The electrolyte is reused after proper purification. The purification incorporates filtration, settling, and/or centrifuge.

Tool-work gap

The gap can, therefore be adjusted by voltage and tool feed rate. Under equilibrium conditions, the tool feed rate (or the material removal rate) is directly proportional to current density. If the feed rate is increased, the electrical resistance of the gap reduces to allow more current to flow.

7.4 DETERMINATION OF METAL REMOVAL RATE

The metal removal rate in ECM process is governed by Michael Faraday's (1791–1867) classical laws of electrolysis.

First law

In the first law of Faraday, it is stated that the amount of chemical change produced by current, *i.e.*, the amount of any substance deposited or dissolved is proportional to the quantity of electricity that is passed through the electrolyte.

Second law

In the second law of Faraday, it is stated that the amount of metal removed from an electrode or deposited on to an electrode by the flow of same quantity of electricity *e.g.*, one Faraday is equal to one gram equivalent weight of the metal.

If W be mass of ions dissolved in gram ion, and Q the quantity of electricity in coulombs,

$$Q = I \times t$$

where I is the current flowing through the electrolytic cell in amp, and t the time in sec. According to Faraday's first law of electrolysis

$$W \propto Q$$

or,
$$W = ZQ$$

or,
$$W = ZIt$$

Where Z is a constant and is known as the electro-chemical equivalent (ECE) of the substance.

If $Q = 1$ coulomb, *i.e.* $I = 1$ amp. and $t = 1$ sec. then,

$$W = Z$$

Hence, ECE of any substance is equal to the mass of the ions liberated by the substance by the passage of one amp. of current for one sec. through the electrolytic solution or by the flow of one coulomb of charge.

From the second law of Faraday, it is found that

$$W = E$$

for the flow of same quantity of electricity *i.e.*, one Faraday.

Here, E = Equivalent weight of a substance dissolved or deposited

$$= \frac{\text{Atomic Weight}}{\text{Valency}}$$

Therefore,
$$E = \frac{M_x}{V_m}$$

where V_m is the valency of metal dissoluted, and M_x the atomic weight of the metal in grams.

The inter-relationship amongst the mass of ions deposited at or liberated from an electrode, their chemical equivalents and their electro-chemical equivalents for two different substances can be shown as :

$$\frac{W_1}{W_2} = \frac{E_1}{E_2} = \frac{Z_1}{Z_2}$$

Where the subscripts 1 and 2 denote two different substances 1 and 2.

The value of Z will be given by,

$$Z = \frac{1}{F}\frac{M_x}{V_m}$$

where, F = Faraday's constant

= 96,500 coulomb

= 26.8 amp. hr.

Hence, with increased valency of an ion, the number of ions required to carry a given quantity of electricity or charge will be less, *i.e.*, the mass of an ion dissolved will be inversely proportional to the valency of the ion.

$$W = ZIt$$

$$= \left(\frac{1}{F}\frac{M_x}{V_m}\right) I \cdot t$$

Thus,

$$W = \frac{M_x I \cdot t}{F V_m}$$

Since,

$$\frac{M_x}{V_m} = E,$$

$$W = \frac{E \cdot I \cdot t}{F}$$

Assuming 100% current efficiency (*i.e.*, all the current flowing through the electrolytic cell is used in the desired metal removal process) in the steady state flow condition, the rate of metal removal, expressed in terms of the height of the removed layer, will be given by

$$MRR = \frac{W}{A\rho t} \text{ cm/sec.}$$

Where A = machined area, cm^2

ρ = density of the workpiece, gm/cm^3

Hence,

$$MRR = \frac{EIt}{FA\rho t} = \frac{EI}{FA\rho}$$

Hence, $$\text{MRR} = \frac{EJ_c}{F\rho} \text{ cm/sec}$$

Where J_c = Current density = I/A

In the actual machining process, the metal removal rate depends on the current efficiency achieved, *i.e.*, the proportion of the total current that is used for the removal of metal from the anode and in practical machining process the whole of the total current is not utilized for machining. The current efficiency is given by

$$\eta_i = \frac{\text{Actual metal removed per amp. min}}{\text{Theoretical amount of metal removed per amp. min}} \times 100\%$$

For multi-valent materials, the current efficiency is influenced by valency of material. Thus, the value of the valency used must be mentioned as in the case of machining iron (Fe^{++} and Fe^{+++}) since the valency may be 2 or 3 in this case. Current efficiency values less than 100% are obtained because of the occurrence of the side reactions at the anode, *e.g.*, evolution of oxygen gas or generation of Fe^{+++} ions instead of Fe^{++} ions.

Dissolution of the metallic ions at a lower valence than desired will cause the current efficiency obtained higher than 100%. For example, the dissolution of iron at a lower valence Fe^{+} than the theoretical Fe^{++} may cause the current efficiency obtained higher than 100%.

7.5 EVALUATION OF METAL REMOVAL RATE OF AN ALLOY

The theoretical metal removal rate of an alloy can be calculated by summing up the charges required for the removal of each of the element from a given volume of the alloy. The analysis for an alloy having the following particulars can be done as described here-in-after

Number of elements = 1, 2, 3, . . ., etc.

Atomic weights of elements = $m_1, m_2, m_3, \ldots$, etc.

Valency at which the elements are dissoluted respectively = $V_1, V_2, V_3, \ldots$, etc.

Density of the alloy = ρ gm/cm^3

Percentage of different elements in order = $X_1, X_2, X_3, \ldots$, etc.

Thus, a volume of the alloy V_A will contain $\frac{V_A\rho X_1}{100}$ gm of element 1, $\frac{V_A\rho X_2}{100}$ gm of element 2, $\frac{V_A\rho X_3}{100}$ gms of element 3, etc. using Faraday's laws of electrolysis, it has been found that

$$W = \frac{M_x \cdot I \cdot t}{FV_m} = \frac{M_x Q}{FV_m}$$

Where, Q is the quantity of electricity or charge;

Hence, $$Q = \frac{WFV_m}{M_x}$$

Thus, the charge required for removing all of the element 1 from the total volume V_A of the alloy is given by

$$Q_1 = \frac{W_1FV_1}{M_1} = \frac{V_A\rho X_1}{100}\cdot\frac{V_1F}{M_1}$$

Similarly, $$Q_2 = \frac{V_A\rho X_2}{100}\cdot\frac{V_2F}{M_2}$$

$$Q_3 = \frac{V_A\rho X_3}{100}\cdot\frac{V_3F}{M_3}$$

and so on.

The total charge required for removing all the elements is given by :

$$Q_{\text{Total}} = Q_1 + Q_2 + Q_3 + \ldots$$

$$= \frac{V_A\rho X_1}{100}\cdot\frac{V_1F}{M_1} + \frac{V_A\rho X_2}{100}\cdot\frac{V_2F}{M_2} + \frac{V_A\rho X_3}{100}\cdot\frac{V_3F}{M_3} + \ldots$$

$$= \frac{V_A\rho F}{100}\left[\frac{X_1V_1}{M_1} + \frac{X_2V_2}{M_2} + \frac{X_3V_3}{M_3} + \ldots\right]$$

Hence, the volumetric metal removal rate per unit of charge is given by

$$V_m = \frac{V_A}{Q_{Total}}$$

$$= \frac{100}{\rho F}\left[\frac{1}{\frac{X_1V_1}{M_1} + \frac{X_2V_2}{M_2} + \frac{X_3V_3}{M_3} + \ldots}\right] \text{cm}^3/\text{amp}/\text{sec.}$$

Example : *In a certain electro-chemical dissolution process of iron, a metal removal rate of 2 cm*^{3}*/min was desired. Determine the amount of current required for the process. Assume :*

Atomic weight of iron = 5.6 gm

Valence at which dissolution occurs = 2

Density of iron = 7.8 gm/cm^3

Solution : Given : atomic weight of iron, M_x = 56 gm.,

Valency of iron dissolution, V_m = 2,

Therefore, grams equivalent weight of iron, $E = \frac{M_x}{V_m} = \frac{56}{2} = 28$

Amount of iron dissoluted, $W = \dfrac{EIt}{F\rho}$

Thus, metal removal rate in cm³/min $= \dfrac{EI}{F\rho}$

Where, $F = 1{,}609$ amp. min

Hence, $2 = \dfrac{28 \times I}{1609 \times 7.80}$

or, $I = \dfrac{2 \times 7.80 \times 1609}{28} = 896$ amp.

7.6 ELECTRO-CHEMISTRY OF ECM

From an analysis of the electro-chemistry of ECM, it has been found that the following two types of reactions usually occur in the system :

(*i*) Electro-chemical reactions at the electrodes, e.g. gas evolution, plating, electrode dissolution and oxidation-reduction.

(*ii*) Chemical reactions in the bulk of the electrolyte, *e.g.*, chemical combinations, complex formation or precipitation reactions.

The electro-chemical reactions occur at the metal electrolyte boundary layers and the transfer of ions in the electrolytic solution takes place by :

(*i*) diffusion

(*ii*) migration in an electrical field

(*iii*) convection in the flow

However, the different electrode reactions takes place as soon as an appropriate potential is reached. Thus, the anode or workpiece having the largest oxidation potential will occur first and the cathode or tool reaction having the smallest oxidation potential will occur last.

Reactions at cathode

The usual types of reaction at the cathode are :

(*i*) plating of metal ions and

(*ii*) evolution of hydrogen gas

The reaction for metal plating is :

$$M^+ + e^- \longrightarrow M$$

Where M represents any metal.

The reactions for hydrogen evolution are :

$$2H^+ + 2e^- \longrightarrow H_2\uparrow \quad \text{(in acid solutions)}$$

$$2H_2O + 2e^- \longrightarrow 2(OH)^- + H_2\uparrow \quad \text{(in alkaline solution)}$$

Reactions at anode *(or Workpiece)*

There will be two types of anode reaction :

(*i*) Dissolution of metal ions in the electrolytic solution.

(*ii*) Evolution of oxygen or halogen gas at the work surface.

The anodic dissolution reaction is :

$$M \longrightarrow M^+ + e^- \text{ (in acid solution)}$$

The oxygen evolution reaction is :

$$2H_2O \longrightarrow O_2\uparrow + 4H^+ + 4e^- \text{ (in acid solution)}$$

$$4(OH)^- \longrightarrow 2H_2O + O_2\uparrow + 4e^- \text{ (in alkaline solution)}$$

The chlorine evolution reaction is :

$$2Cl \longrightarrow Cl_2 + 2e^-$$

polarisation makes the process complicated and hence an additional voltage in excess of the theoretical voltage is necessary to overcome electrode polarization. This excess voltage is called "overpotential" and is classified into the following three groups :

1. **Ohmic potential (E_R)**

   It arises out of the resistance offered to the passage of current, *e.g.*, due to the oxide film formed on the electrode.

2. **Activation overpotential (E_A)**

   It is an additional potential necessary for maintaining the slow electrode reactions at the desired rate. This can be obtained by Tafel relationship given by :

$$E_A = a' + b' \log j$$

where E_A = activation overpotential

j = current density

a', b' = constants for the particular reaction.

3. **Concentration overpotential (E_c)**

   This occurs owing to the difference in concentration at the electrode-solution interface and in the bulk of the electrolytic solution. Hence, standard electrode potentials must be corrected for concentration according to the equation :

$$E_c = E_o + \frac{R\theta}{V_m F} \log \frac{\text{Concentration of reactants}}{\text{Concentration of products}}$$

where, E_c = Corrected potential

E_o = Oxidation potential

R = Universal gas constant

θ = Absolute temperature

V_m = Valency of ion

F = Faraday's constant

The total overpotential E_{Ta} at the anode for a given current density is given by :

$$E_{Ta} = E_o + E_{Aa} + E_{ca} + E_{Ra}$$

Similarly, taking into account the total overpotentials at the cathode, the total voltage (V) across the ECM cell will be given by :

$$V = E_c + E_{Aa} + E_{ca} + E_{Ra} + E_{oc} + E_{Ac} + E_{Cc} + E_{Rc} + I^2R_c + E_{Bc}$$

where, E_{oc} = Cathode reversible potential

I = Current flowing through the cell

R_c = Electrolyte resistance

E_{Bc} = Voltage drop due to attached bubble layer.

7.7 PROCESS PARAMETERS

The metal removal in ECM process is primarily controlled by current density. ECM machines are available that deliver currents from 50 to 40,000 amp. The machine used for a particular application must have sufficient current available to maintain a current density of 8–233 amp/cm^2 at the workpiece. When all other variables are held constant, the tool penetration rate is directly proportional to the current density. Typical ECM metal removal rate is 1.6 cm^3 per minute for every 1000 amp. of current.

Current density is controlled not only by the amount of current that the power supply is delivering, but also by the size of the gap between the tool and the workpiece. A small gap results in the highest current density. The gap may be as small as 0.025 mm or as large as 0.76 mm, the gap size most often used being 0.25 mm. When the gap is very small, there is a danger of sludge particles bridging the gap between the tool and workpiece and causing a short circuit. When the gap is too large, current density is reduced, resulting in a poor surface finish and decrease in material removal rate.

The flow velocity of electrolyte through the gap is also an important parameter affecting the surface finish and removal rate. Under normal operation, the electrolyte flow may be between 15 and 60 m/sec. If electrolyte velocity is too low, the heat and by-products of the reaction (hydrogen gas bubbles and sludge) build in the gap causing non-uniform material removal. A very high velocity will cause cavitation, also promoting uneven material removal. In general electrolyte flow is 0.95 litre/min in case of sodium chloride.

Electrolyte pressure and temperature also play an important role in controlling the process. The electrolyte pressure is dependent upon the flow rate and varies between 69 KPa and 2.7 MPa, depending upon the application. The temperature of the electrolyte can be between 24 and 65°C but the selected temperature must be maintained within a few degrees because electrical conductivity and process repeatability are affected by these changes.

The feed rate varies between 0.5 and 19 mm/min. The amount of overcut that occurs at the sides of the ECM tool is dependent upon the feed rate. A low feed rate will produce a large overcut, and conversely, a high feed rate will reduce the amount of overcut. The interesting relationship between various parameters are, as the current density is increased, the feed rate increases (to maintain a constant cutting gap), the overcut is reduced, and the surface finish is improved.

7.8 PROCESS CAPABILITIES

The metal removal rate in ECM process for various pure metals are given in table 7.2.

Under ideal conditions, ECM is capable of producing tolerances of approximately ± 0.012 mm, although in daily production the number is closer to ± 0.05 mm.

TABLE 7.2. ECM Removal Rate

Material	Density (g/cc)	Removal Rate g/1000 amp. hr	Removal Rate cc/min
Aluminium	2.7	299	1.9
Beryllium	1.8	150	1.3
Cobalt	8.9	978	1.9
Copper	8.9	2129	3.9
Iron	7.8	937	2.1
Lead	11.3	3479	5.0
Molybdenum	10.1	1069	1.8
Nickel	8.9	978	1.9
Titanium	4.5	534	1.9
Tungsten	19.3	1023	0.8
Uranium	19.0	1997	1.8
Zirconium	6.5	761	1.9

The typical overcut at the side of tools is approximately 0.12 mm. Depending upon the tool design being used, taper can be held to 0.025 mm/mm. Surface finish is dependent upon the workpiece material, the type of tool used, the electrolyte flow, and the current density. Usually the surface finish at the tip of an ECM tool will be 0.1–1.5 μ. The side of the tools may produce surface finishes as rough as 5 μ because of low current density.

Using hollow tubing as the ECM tool, holes as small as 0.76 mm in diameter can be drilled. Length/diameter ratios upto 20:1 can be accomplished.

No detrimental effects on materials have been found with ECM when parameters are properly selected.

When compared with mechanical milling, a part produced by ECM will have slightly less fatigue resistance than a part produced by a conventional mechanical process because the ECM process removes material without inducing any residual stresses in the part surface. Mechanical milling, on the other hand, often leaves the part surface in a favourable state of compressive residual stress. Shot peening can be used to improve the fatigue properties of ECM-processed hardware by creating a compressive residual stress in the part surface.

Some specific application of ECM are for (*i*) facing and turning complex three-dimensional surfaces, (*ii*) die sinking, particularly deep narrow slots and holes (*iii*) profiling and any odd shape contouring, (*iv*) multiple hole drilling (*v*) trepanning (*vi*) broaching, (*vii*) deburring, (*viii*) grinding, (*ix*) honing etc.

7.9 ELECTROCHEMICAL DEBURRING

Electrochemical deburring (ECD) is a process for the removal of metal burrs by anodic dissolution, the same principle as ECM. Electrochemical deburring is used exclusively to deburr or radius workpieces. ECD differs from ECM in that the electrolyte pressure, electrolyte flow and current are all relatively low. Another major difference is that the cathode (tool) is usually held stationary in ECD.

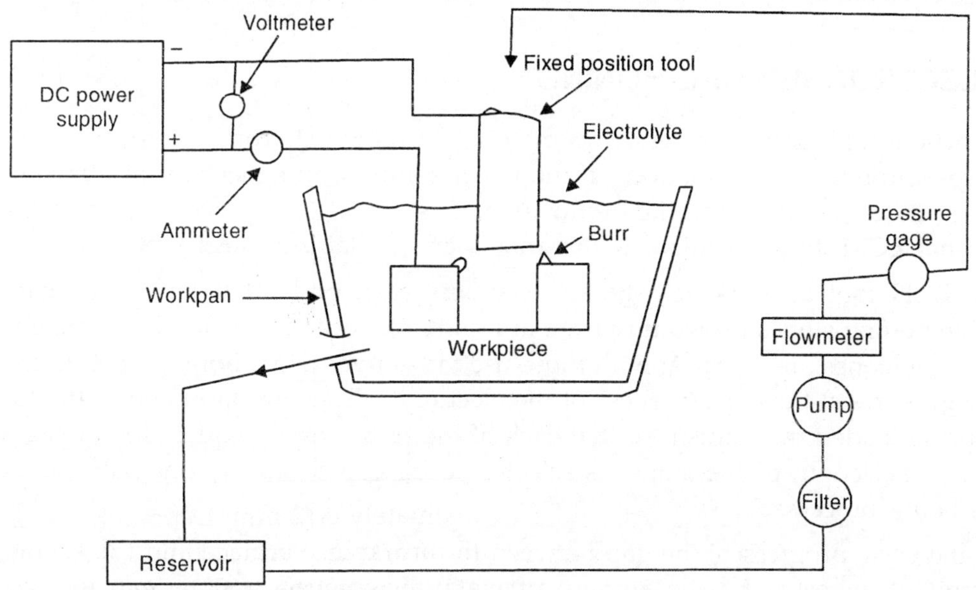

Fig. 7.4

The ECD tool is insulated on all surfaces except those adjacent to the burrs. The tool is positioned to concentrate the deplating action at the root of the burrs. As in ECM, an electrolyte is introduced through the gap between the burr and the tool.

Process parameters

In ECD process, the current ratings ranges from 100 to 2000 amp. at a voltage of 7–25 VDC. The electrolyte pressure ranges from 69 to 345 KPa at a flow rate of 3.7 to 15L/min for each 100 amp. The gap between the tool and the burr is typically 0.76 mm but can range from 0.12 to 1.2 mm. Depending upon the size and geometry, burrs are quickly dissolved in 5–50 sec. leaving a surface finish of 0.2–1.6 μ. Since burr size can vary from part to part, the parameters should be set to remove the largest anticipated burr.

Application examples

The process should not be applied for applications requiring burr removal over large surfaces that would best be processed by mass finishing. Instead, ECD should be applied to selectively-deburr small areas or areas that are impossible to access by other means.

The ECD is not used for long or large burrs, typically created by dull cutting tools. Burr size must be reasonably repeatable from one part to the next otherwise incomplete burr removal or short circuits can result.

The ECD process has been successfully applied in many different industries ranging from consumer appliances and automotive to biomedical and aerospace products.

The ECD process is used in automotive industry for deburring and radiusing the cross-holes in cramkshafts.

7.10 ELECTROCHEMICAL HONING

In electrochemical honing process, the material is removed from electrically conductive workpieces through a combination of anodic dissolution and mechanical abrasion. Eight percent or more, of the material removal occurs through electrolytic action. As with conventional ECM, the workpiece is the anode, and a stainless steel tool is the cathode.

The ECH tool consists of a hollow stainless steel body that has expandable, non-conductive honing stones protruding from at least three locations around the circumference. These honing stones are identical with those used in conventional honing operations, except that they must resist the corrosiveness of the electrolyte. They are mounted in the tool body with a spring-loaded mechanism so that each of the stones exerts equal pressure against the workpiece. The length of the stones is selected to be approximately one-half the length of the bore being processed.

As the cycle progresses, the gap between the workpiece surface and the stainless steel tube (cathode) increases by the amount of material removed. As the gap increases, and electrochemical action slows, a proportionately larger percentage of the material is removed by mechanical abrasion.

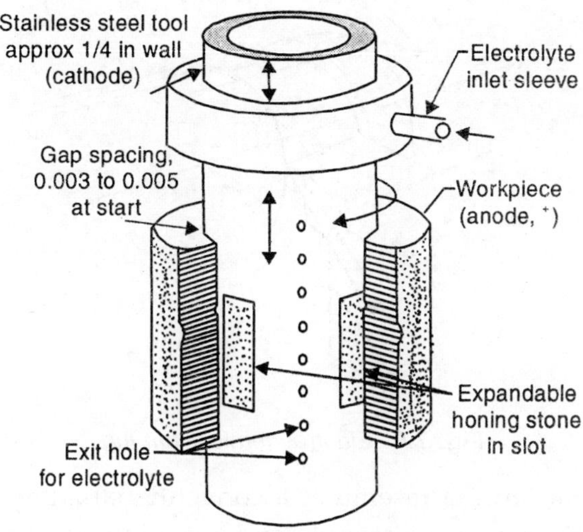

Fig. 7.5. *Electrochemical honing*

Material of any hardness can be machined by ECH, as long as it is electrically conductive. It can remove material at rates upto 100% faster than conventional honing, the gain being more pronounced as the material hardness increases. As most of the material is removed electrolytically, no microscratches are left on the work surface as it happens in case of conventional honing. ECH imparts no residual stresses in the workpiece.

Size tolerance of 0.01 mm on the diameter can be obtained and roundness can be maintained at less than 0.005 mm. Surface roughness of the order of 0.1–0.5 micron CLA is obtainable by this method.

Application

The process is easily adaptable to cylindrical parts for trueing the inside surfaces. The size of the cylinder that can be processed by this method is limited only by the current and electrolyte that can be supplied and distributed.

7.11 ELECTROCHEMICAL TURNING

The electrochemical machining process can be used for turning operation. Electrochemical turning remove the material from the internal surface (ID), external surface (OD), or face of rotating electrically conductive workpieces.

The tool of ECT can be constructed as wide as the final shape being generated if machine is of sufficient power. By this process, contoured shapes can be plunged into the workpiece in a single pass. During operation, a gap is constantly maintained between the workpiece and tool and electrolyte is flooded continuously in the gap. The electrolyte is delivered through holes in the tip of the tool.

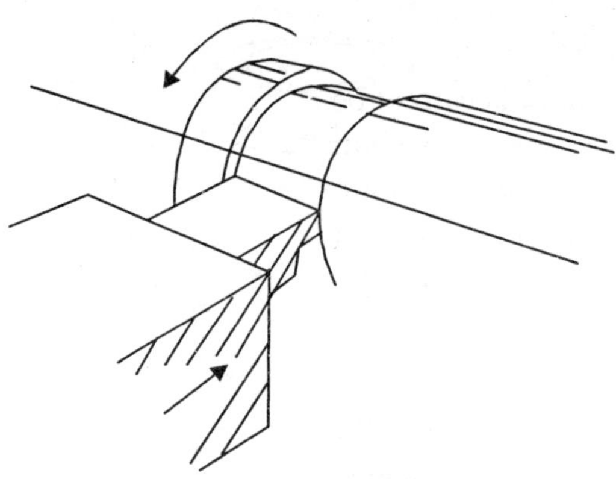

Fig. 7.6. *Electrochemical turning*

The equipment in the process resembles a conventional lathe except for the additional enclosures to contain the corrosive electrolyte and the addition of the power supply. Power supplies have been attached to Electrochemical turning machines with ratings as high as 20000 amp.

Very thin workpiece sections can be turned or faced with ECT because there is no physical contact between the ECT tool and workpiece. Disks as thin as 0.12 mm can be turned without distortion because there are no tool forces. In addition, material hardness has no influence on removal rates and there is no tool wear to decrease quality or productivity.

The ECT process is capable of tolerances as close as ± 0.007 – ± 0.012 mm. Surface finish can be 0.12μ.

At present ECT are centred around the aerospace industry such as turning turbine wheels, rough-turning large disk forgings and turning hardened bearing races.

7.12 PROCESS SUMMARY

Advantages

One of the main advantages of ECM lies in its ability to machine complex three-dimensional curved surfaces without the striation marks left by milling cutters. By preparing one set of cathode tools, say, from stainless steel, it is possible to machine many roughly formed components to within close limits of the desired contour. For this reason, gas and steam turbine blades are now machined by this method.

The process is capable of machining metals and alloys irrespective of their strength and hardness. ECM offers a higher rate of material removal in case of hard material as compared with conventional machining methods.

Limitations

The non-conductive materials cannot be machined. This method is incapable to machine sharp interior edges and corners (less than 0.2 mm radius) because of very high current densities at those points. Blind holes cannot be machined in the solid block in one stage. Corrosion and rust of the ECM machine can be a hazard. But preventive measures can help in this regard. Less corrosive though slightly more expensive electrolytes, like sodium nitrate, can also be employed.

Although the parts produced by ECM are stress-free, they are found to have fatigue strength or endurance limit lowered by approximately 10 to 25 per cent.

Corrosion and rust of the ECM machine can be a hazard. To prevent it the interior parts may be coated by heat cured polyethane; high intensity spray areas by keroseal and brine tank may lined by vinyl.

More space is required in comparison to conventional machining methods. Some additional problems are related to machine tool requirements, such as power supply, electrolyte handling and tool feed servo systems.

REVIEW QUESTIONS

1. Explain the working principle of electro-chemical machining process. Give a list of the hardware used in this process.
2. Explain how would you make use of the Faraday's law of electrolysis for computing the material removal rate and the tool feed required during electrochemical machining.
3. What are the typical functions expected to be served by an electrolyte in ECM process ? Name a few important electrolytes.
4. What factors need to be considered while selecting the electrolyte ? Discuss the advantages and limitations of some electrolytes.
5. (*i*) Describe the chemistry involved in the ECM process.

   (*ii*) What is the "self adjusting feature" in ECM ?
6. Write short noes on :

   (*i*) The economics of electrochemical machining.

   (*ii*) The effect of high temperature and pressure of electrolyte in the ECM process.

   (*iii*) Applications of electrolytic grinding process.
7. While machining tungsten, a metal removal rate of 0.98 cm/min has been observed at 1000 amp. Evaluate the valency at which the metal has been electro-chemically dissoluted. Assume

   (*i*) Atomic weight of tungsten = 186 gm, and

   (*ii*) Density of tungsten = 19.4 gm/cm^3

8. Calculate the metal removal rate (m^3/min) in an anodic dissolution process of chromium if the current density available has been 250 amp/cm^2. Assume

atomic wt. of chromium = 52 gm

valency of chromium at which dissoluted = 2

density of chromium = 7.2 gm/cm^3

9. During electro-chemical machining of iron with a copper electrode working in a 5(N) NaCl solution in water, an equilibrium gap of 0.0125 cm has been achieved at a current density of 150 amp/cm^2. If the operating voltage has been 10 V, determine the feed rate of the tool. Given :

(*i*) Specific resistance of NaCl solution = 3 ohm-cm

(*ii*) Valency at which iron is dissoluted = 2

(*iii*) Density of iron = 7.8 gm/cm^3.

Electro-chemical Grinding

8.0 INTRODUCTION

The machining of carbides and other hard to machine materials has been limited to the diamond wheel grinding for a long time. But the process has become costly because of the scarcity and high cost of the abrasives necessary for the diamond wheel. The electro-chemical grinding is developed to grind more effectively and economically cemented carbides and other current conducting material which are difficult to machine. In the early days of the development of ECG, the process was thought as "electrolytically assisted grinding." But, with the development of the process, it has been found that 95% of the metal is removed by electrolytic action while machining by ECG process employing diamond wheel. Hence, it may be called "mechanically assisted electrochemical machining". The wheels usually used in ECG process, namely, diamond wheels for cemented carbides and conventional abrasive wheels for steels, are metal bonded.

8.1 FUNDAMENTAL PRINCIPLES OF ECG

Metal removal by ECG process comprises :

(*i*) a mechanical abrasive action,

(*ii*) the anodic dissolution of the workpiece, and

(*iii*) the removal of oxide films.

This process employs a grinding wheel in which an insulating abrasive is set in a conducting bonding material. The D.C. power connected to the part and the conductive bond of the grinding wheel in such a way that the latter is at negative potential with respect to the component part. The electrolyte, usually a neutral salt solution such as sodium chloride, sodium nitrate, sodium nitrite, potassium nitrate, is passed through the machining zone in order to complete the electrical bridge between the anode and cathode as shown in figure 8.1.

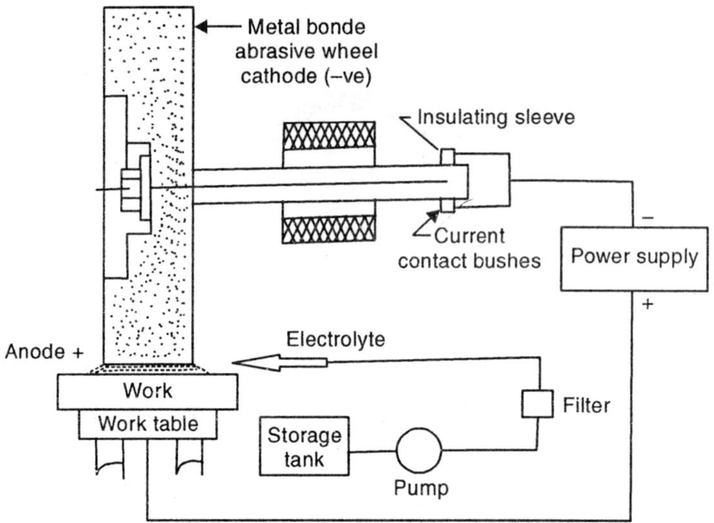

Fig. 8.1 *Schematic of electrolytic grinding process*

The abrasive particles in an ECG wheel protrude beyond the conductive bond surface. This establishes a small gap between the wheel and workpiece. Electrolytic action begins when the gap is filled with an electrolyte and the wheel is electrically charged. Material is removed through a combination of electrochemical action and conventional mechanical grinding. Approximately 90% of the material is removed through electrochemical action and 10% by mechanical grinding.

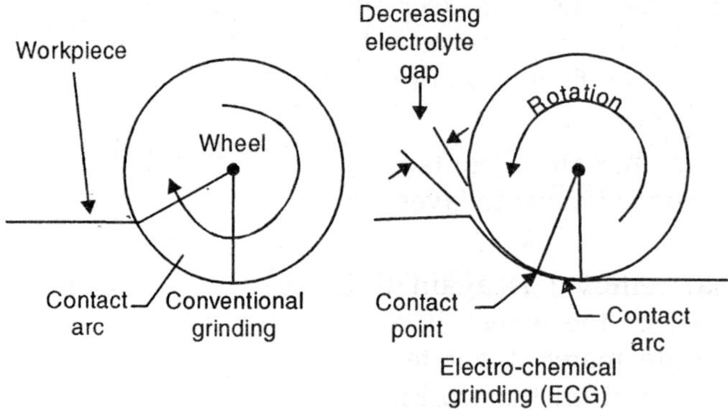

Fig. 8.2. *A small mechanical contact arc is partly responsible for the long wheel life experienced by ECG users*

Because only a small amount of material is removed by grinding action, ECG wheel life is typically 10 times longer than the life of a conventional grinding wheel. An additional factor contributing to the long life of ECG wheels is the contact arc. The mechanical contact arc for the ECG process is only a fraction of that used for conventional grinding.

The mechanical contact arc of an ECG wheel is smaller than that of a conventional grinding wheel because ECG material removal occurs in three phases corresponding to three zones on the wheel. Phase–1 is entirely electro-chemical and takes place within a zone at the leading edge of the ECG wheel as shown in figure 8.3.

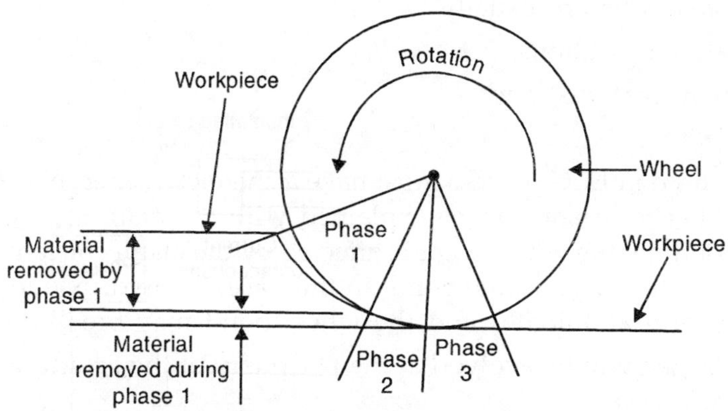

Fig. 8.3. *The three phases of ECG material removal*

The fresh electrolyte from delivery nozzle is allowed to move into and through the wheel-workpiece interaction zone. As electrolyte passes through the phase–1 zone and moves toward the phase–2 zone, oxides, evolved gases and numerous other process by products contaminate the electrolyte and decrease its conductivity. As the conductivity decreases, less and less of the workpiece is depleted, causing the gap to decrease until finally the wheel comes into contact with the workpiece. At this point on the wheel contact arc, phase–1 material removal ends and phase–2 begins.

Phase–2 material removal begins at the point where the abrasives in the ECG wheel are in direct contact with the workpiece. This traps the electrolyte between the protruding abrasive grains, thus forming tiny electrolytic cells where the depleting occurs. In addition, because the electrolyte is being forced deep into the contact arc by the rotational motion of the wheel, the local electrolyte pressure in the gap increases. Increasing the electrolyte pressure acts to suppress the formation of gap bubbles in the gap, which in turn increases the material removal rate.

The abrasive grains that are in contact with the workpiece surface during phase–2 act to remove the soft, non-reactive oxide layer, thus exposing fresh metal to further electrolytic action.

Phase–3 material removal is again totally electrochemical. As the contaminated electrolyte exits the rear of the wheel, it continues to slowly deplete the workpiece. This phase removes very little material but tends to electrolytically remove any scratches or burrs that may have formed on the workpiece.

8.2 ELECTRO-CHEMISTRY OF ECG

The electro-chemistry of ECG reveals that :

(*i*) electro-chemical reactions occur at the metal-electrolyte boundary layer, and

(*ii*) chemical reactions occur in the bulk of the electrolyte solution.

The electro-chemical reactions comprise

(*i*) anodic dissolution,

(*ii*) gas evolution at the electrodes, and

(*iii*) oxidation-reduction, etc.

The chemical reactions are usually

(*i*) chemical combinations,

(*ii*) complex formations, and

(*iii*) precipitation.

The electrolyte used in ECG process must have a chemically inert effect on the conductive wheel bond material. The reason can be explained with the help of an example. In case of aluminium based bonding, the wheel wear is affected by the alkali concentration, for example hydroxyl ion concentration around the bond. In this case, the hydroxyl ions chemically react with aluminium to form aluminates and the wheel bond may break down.

Gases like hydrogen will be evolved at the periphery of the grinding wheel while using a neutral electrolyte.

At the anode :

$$M = \text{Workpiece of any metal}$$

$$M \rightarrow M^+ + e^-$$

$$M^+ + OH^- \rightarrow MOH + e^-$$

$$2H_2O \rightarrow O_2 + 4H^+ + 4e^- \text{ (in acid solution)}$$

$$4(OH)^- \rightarrow 2H_2O + O_2\uparrow + 4e^- \text{ (in alkali solution)}$$

At the cathode :

$$2H^+ + 2e^- \rightarrow H_2\uparrow \text{ (in acid solution)}$$

$$2H_2O + 2e^- \rightarrow 2(OH)^- + H_2\uparrow \text{ (in alkaline solution)}$$

8.3 PROCESS PARAMETERS

The effectiveness of ECG process depends on the various process parameters such as current density, voltage, electrolyte temperature, feed rate, gap between work surface and the conductive wheel etc.

The metal removal rate for ECG is governed by the current density. Standard machines are available with current ratings ranging from 50 to 3000 amp. The current density most often used depends upon the material being processed and can range from 77 to 620 amp/cm^2. As with ECM, the higher the current density, the faster the removal rate and the better the resulting surface finish.

The voltage used for ECG can range from 4 to 15 VDC depending upon the application and is usually set as high as possible to maximise the removal rate. The maximum limit must not be so high that it degrades the quality of the machined surface, and it must be consistent with surface finish and tool requirements. Typically, voltage is adjusted by increasing it until fine red sparks become visible at the exit side of the wheel. Too much voltage is indicated by the formation of bright blue, audible sparks at the front of the wheel.

The electrolyte is maintained at a temperature between 15 and 32°C and is filtered as in other ECM processes. The electrolyte being forced through the gap generates a wheel pressure. In practice, this pressure can range from 0.3 to 1.4 MPa and is held constant by either a constant feed rate or a hydraulic cylinder that allows the wheel to either give or feed depending upon the dissolution rate.

Once the feed rate is established for a particular application, it can be varied by as much as 4 mm/sec. If the feed rate is running too slowly for the application, a large overcut will be produced that will result in poor surface finishes and tolerances. If the feed rate is too fast, the abrasive particles will be prematurely forced into the workpiece, resulting in excessive wheel wear.

The gap between the work surface and the conductive wheel bonding agent is normally 0.25 mm on a freshly dressed wheel. Wheel speeds are usually between 1200 and 1800 surface meters per minute.

The maximum depth of cut for a single ECG wheel pass is usually considered to be about 2.5 mm but can be extended with care. The limiting factor for this is the wheel contact arc, which should never exceed 19 mm. When the wheel contact arc extends beyond this limit, the electrolyte becomes ineffective because of the high concentration of hydrogen bubbles and dissolved metal oxides.

8.4 PROCESS CAPABILITIES

Electrochemical grinding exhibits material removal rates that are upto 10 times faster than conventional grinding on materials harder than HRc 60. It is considered cost effective on materials harder than HRc 45. Material removal rates average 1.6 cm^3 with the range being from 0.6 to 5 cm^3.

Although removal rates are high, ECG cannot obtain conventional grinding tolerances. The current limits are ± 0.025 mm using standard techniques and ± 0.012 mm when a no current finishing pass is used. The minimum radius that can be produced on an inside corner is 0.25 mm; outside corners can be produced with a radius smaller than 0.025 mm.

Surface finishes average 0.2–0.8μ, with plunge-grinding methods giving a 0.12–0.24μ finish, and surface grinding producing finishes of 0.24–0.37μ. No matter how hard, tough, or brittle a material is, ECG produces surfaces free of any grinding scratches, burrs, or burns. Electrochemical grinding will also produce a better surface finish than conventional grinding in non-homogeneous materials.

When grinding tough alloys, ECG metal removal rates may be five to ten times those of conventional grinding methods. After taking wheel wear into consideration, total production has been reported to be 20 times that of conventional grinding.

Table 8.1 below lists the operating conditions and results for various materials processed by electrochemical grinding.

TABLE 8.1. ECG Operational Examples

Material	Wheel width (mm)	Volts	Current (amp)	Depth of cut (mm)	Feed (mm/min)
M2 tool steel	12.7	10	225	2.5	17
U–700	76.0	6	2700	3.8	38
Waspaloy	12.7	6	450	0.8	64
Tungsten	2.0	8	150	1.0	19
Stellite	12.3	7	60	9.5	52

8.5 APPLICATION EXAMPLES

The primary application for ECG is grinding carbide cutting tool inserts. In this case, an 80% savings in wheel costs and 50% savings in labour can be realised when compared with conventional grinding.

The ECG is used to reprofile worn locomotive traction motor gears. In this application, large, hardened gear teeth are ground to remove all indications of wear from the tooth surfaces, thus re-establishing a new tooth contour. Often as much as 0.38 mm of material is removed for this purpose.

ECG is used for this purpose because it is unaffected by material hardness, it does not alter the gear hardness or residual stresses and it does not require frequent wheel dressing.

Other applications of ECG involve the burr-free sharpening of hypodermic needles, grinding of superalloy turbine blades, and form grinding of fragile aerospace honeycomb metals.

End mills are ground with a full-depth single pass for each flute as compared with the maximum conventional removal rate of 0.025 mm per pass that is required to avoid burning.

8.6 PROCESS SUMMARY

Advantages

(*i*) Increased material removal rates.

(*ii*) Reduced cost of grinding. Despite the fact that the machine required is comparatively expensive, the increased rate of metal removal and the reduced consumption of abrasive material more than compensate for the extra capital cost.

(*iii*) Reduced heating of workpiece, and therefore, less risk of thermal damage.

(*iv*) Absence of burrs on the finished surface.

(*v*) Improved surface finish with no grinding scratches.

(*vi*) Reduced pressure of work against the wheel.

Disadvantages

(*i*) High capital cost.

(*ii*) Corrosive environment.

(*iii*) High preventative maintenance cost.

REVIEW QUESTIONS

1. How Electro-chemical grinding differs from the conventional grinding operations ?
2. Explain the mechanism of metal removal in ECG process.
3. What are the different process parameters considered for evaluation of effectiveness of ECG.
4. Brief the major area of application of ECG process.

❑❑❑

Electrical Discharge Machining (EDM)

9.0 INTRODUCTION

The development and usage of materials like high-strength-temperature-resistant alloys and other hard to-machine conducting materials has paved the ways for the development of some non-traditional techniques of machining utilizing the thermo-electric source of energy. Electrical discharge machining is one of such machining processes which has been of immense help to the manufacturing and processing engineers to produce intricate shapes, on any conducting metal and alloy irrespective of its hardness or toughness.

In 1943 in USSR, Lazarenko and Lazarenko developed the construction of the first spark erosion machine. He decided to use the destructive effect of an electrical discharge and developed a controlled method of metal machining. The Lazarenko circuit, used for power supplies for EDM machines in 1943 is improved to present form being used in many current applications. EDM has found application in the machining of hard metals or alloys which cannot be machined easily by conventional methods. It plays a major role in the machining of dies, tools, etc. made of tungsten carbides, stellites or hard steels. Alloys used in aerospace industries, nuclear power plant such as hastalloy, nimonic etc. could also be machined conveniently by this process.

The basic scheme of electric discharge machining is shown in figure 9.1.

The machining process involves controlled erosion of electrically conducting materials by the initiation of rapid and repetitive electrical spark discharge between the tool (or cathode) and the workpiece (or anode) separated by a dielectric fluid medium. When a discharge takes place between two points of the anode and the cathode, the intense heat generated near the zone melts and evaporates the materials in the sparking zone. For effective machining, the workpiece and the tool are submerged in a dielectric fluid (hydrocarbon or mineral oils). It is observed that erosion of positive terminal is faster than that of negative terminal. For this reason, the workpiece is normally made the anode. A suitable gap, known as spark gap, is maintained between the tool and the workpiece to cause spark discharge. The spark gap can be varied to match the machining conditions such as metal removal rate. The sparks are made to discharge at a high frequency with a suitable source. Since the spark occurs at the spot where the tool and the workpiece surfaces are the closest and since the spot changes after each spark (because of the material removal after each spark), the sparks travel all over the surface. This results in a uniform material removal all over the

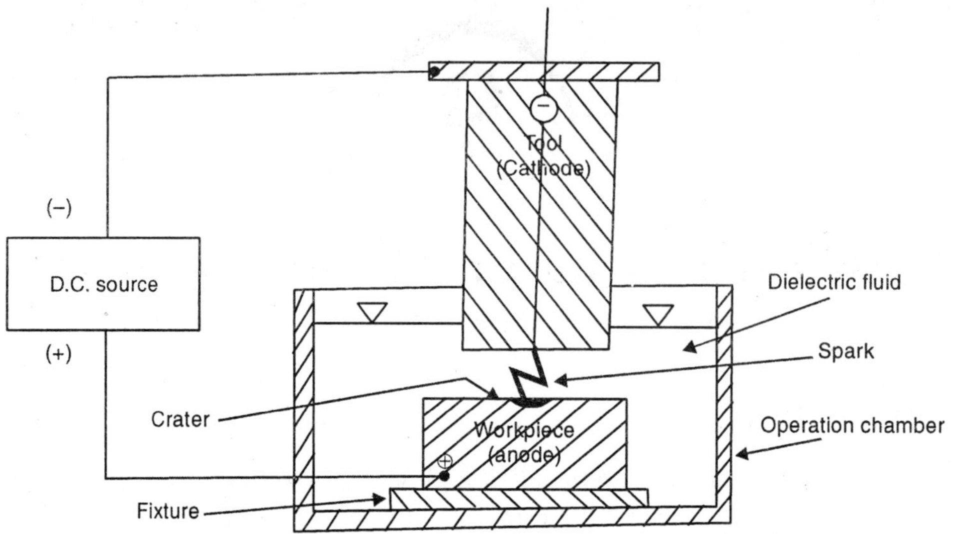

Fig. 9.1. *Basic scheme of EDM process*

surface, and finally the work face conforms to the tool surface. This way, the tool produces the required impression in the workpiece. For maintaining the predetermined spark gap, a servocontrol unit is generally used. The gap is sensed through the average voltage across it and this voltage is compared with a preset value. The difference is used to control the servomotor. Sometimes, a stepper motor is used instead of a servomotor. The spark frequency is normally in the range 200–500000 Hz, the spark gap being of the order of 0.025–0.05 mm. The peak voltage across the gap is kept in the range 30–250 Volts. A metal removal rate upto 300 mm^3/min can be obtained with this process, the specific power being of the order of 10W/mm^3/min. The efficiency and the accuracy are improved when a forced circulation of the dielectric fluid is provided. The commonly used dielectric fluid is kerosene. The tool is generally made of brass or a copper alloy.

9.1 CLASSIFICATION OF ELECTRIC DISCHARGE MACHINING

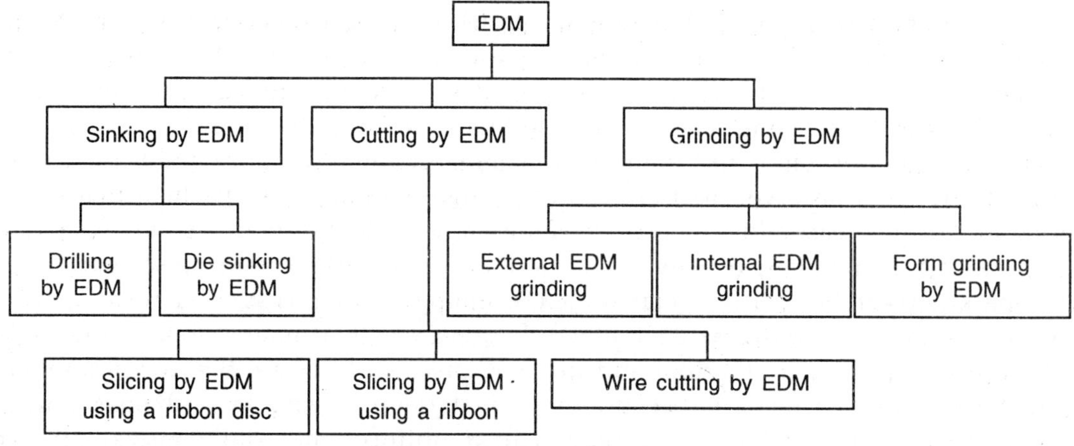

Fig. 9.2

Sinking by EDM

In this process, non-stationary electrical discharges are used for the metal removal which are separated from each other spatially and temporarily. This process involves those EDM operations in which the average relative speed between the tool and work piece is coincident with the penetration speed in the workpiece.

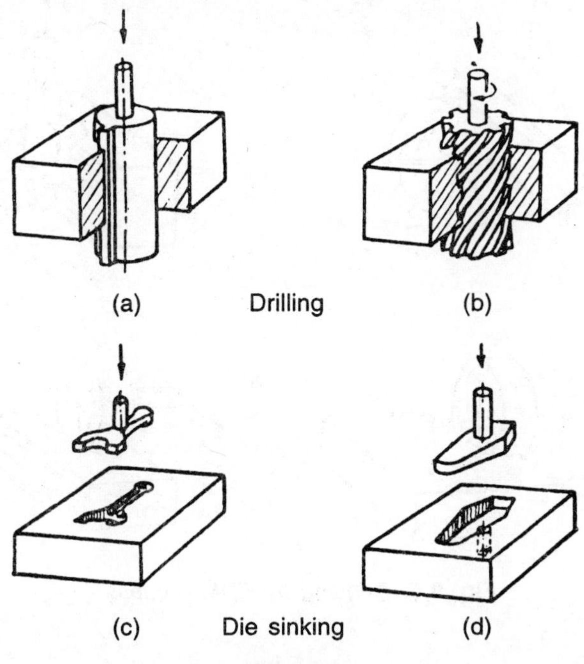

Fig. 9.3

Cutting by EDM

This process includes the machining operations where the workpiece is cut-off or notched. The figure 9.4, demonstrate the various EDM cutting operations.

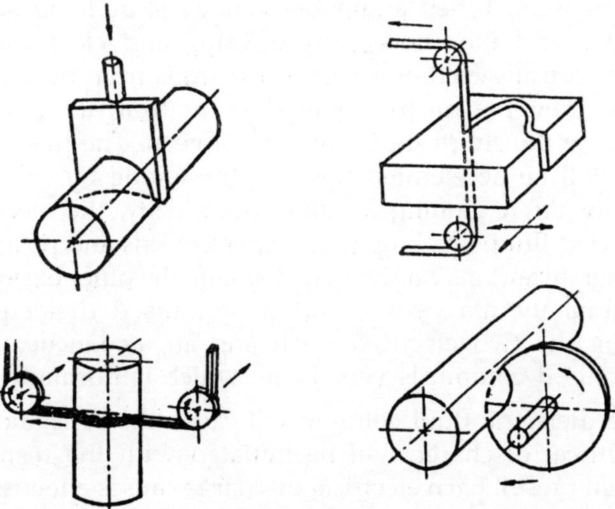

Fig. 9.4

Grinding by EDM

This process incorporates the machining operations made with an electrode rotating around an axis in addition to the normal electrode feed. The details arrangements for this process are shown in figure 9.5.

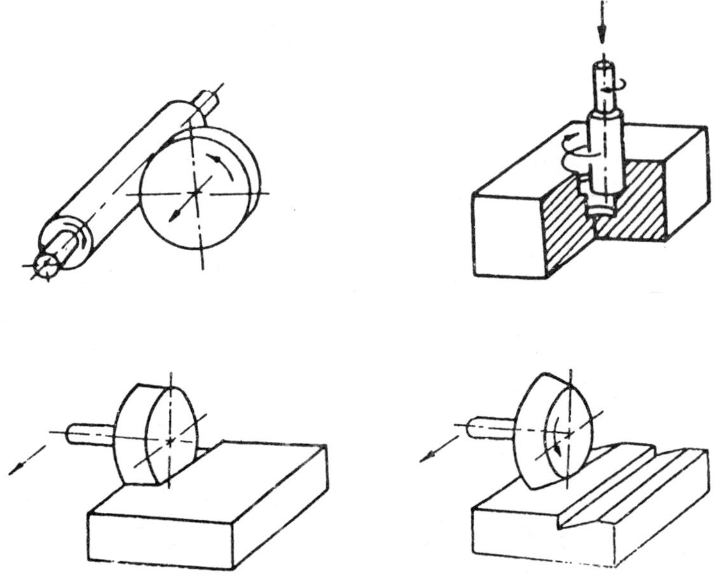

Fig. 9.5. *Grinding by EDM process*

9.2 MECHANISM OF METAL REMOVAL

In single discharge of EDM process, the sequence of events take place for removing the metal from workpiece are explained as follows :

Consider the case of a discharge between two electrodes (tool and workpiece) through a gaseous or liquid medium. When a suitable voltage is built up across the tool and the workpiece (the cathode and the anode, respectively), and electrostatic field of sufficient strength is established, causing cold emission of electrons from the cathode surface towards the anode originating from the micro-irregularities or from the small protrusions on the electrode surfaces having the shortest distance in between. The free electron, liberated from the cathode surface, will be accelerated towards the anode by the electric field and will acquire a high velocity. After gaining a sufficient velocity, the electrons collide with the molecules of the dielectric fluid, breaking them into electrons and positive ions. The electrons so produced also accelerate and may ultimately dislodge the other electrons from the dielectric fluid molecules. Ultimately, a narrow column of ionised dielectric fluid molecules is established connecting the two electrodes (causing an avalanche of electrons, since the conductivity of the ionised column is very large, which is normally seen as a spark).

Ionisation of the dielectric fluid column will cause the resistance of the fluid layer to decrease and an electrical discharge will be initiated with the resulting flow of electron (which yield the current pulse). Each electrical discharge causes a focussed stream of electrons to move with a very high velocity and acceleration from the cathode towards the anode and ultimately creates compression shock waves on both the electrode surfaces.

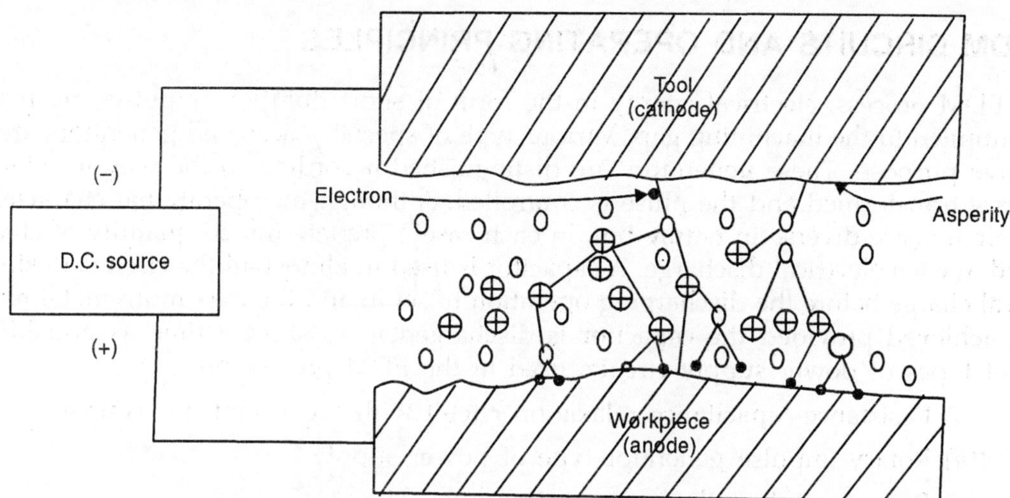

Fig. 9.6. *Scheme of dielectric breakdown*

The pressure created by the waves is many times higher than the ultimate strength of the electrodes. The generation of the compression shock waves develops a local rise in temperature and a certain surface layer deformation. The phenomenon is accomplished within a few microseconds and the temperature of the spot hit by electrons may rise upto 12000°C causing fusion or partial vaporisation of the metal and the dielectric fluid at the point of discharge. The metal in the form of liquid drops is dispersed into the space surrounding the electrodes by the explosive pressure of the gaseous products in the discharge. This results in the formation of a tiny crater at the point of discharge in the workpiece. Comparatively less metal is eroded from the cathode (tool) as compared to the anode (work) due to the following reasons :

(*i*) A compressive force is created on the cathode surface by the spark and thus produces reduced cathode wear.

(*ii*) The momentum with which the positive ions strike the cathode surface is much less than the momentum with which the electron stream impinges on to the anode surface, though the mass of an electron is much smaller than that of a positive ion. This is due to the fact that the striking velocity of electrons is much higher than that of positive ions.

(*iii*) Due to pyrolysis of dielectric fluid, gases like 60–70% hydrogen, 15–25% acetylene and the rest paraffin hydrocarbons are evolved. The carbon particles produced from these gases are deposited on the heated cathode surface as a thin film of crystalline graphite and lessens the cathode wear.

The spark is initiated first in the zone where the distance between the tool and the work surface is the smallest. The crater is formed on the tool and workpiece surface which led to increase in gap in the shortest distance region. This causes the spark to shift to other zone having the minimum spark gap. This process continues till an even spark gap is maintained throughout the entire work surface for particular tool-work setting.

9.3 EDM CIRCUITS AND OPERATING PRINCIPLES

In the EDM process, electrical energy in the form of short duration impulses are required to be supplied to the machining gap. Various type of specially designed generators are used for above purpose. These generators are distinguished according to the way in which the voltage is transformed and the pulse is controlled. Although the operational characteristics of the circuits are diverse in nature but, in each case a predetermined quantity of charge is released at each electrical discharge. A capacitor is used in almost all the circuits to store the electrical charge before the discharging operation to set in and the maximum metal removal rate is achieved provided the capacitor is discharged in as short a time as possible. The different types of power supply circuits used in the EDM process are :

(*i*) Resistance-capacitance relaxation circuit with a constant d.c. source.

(*ii*) Rotary impulse generator type of power supply.

(*iii*) Power supply utilizing controlled pulse circuits.

Resistance-capacitance relaxation circuit

The resistance-capacitance relaxation circuit was used when the electric discharge machines were first developed. In this circuit, the condenser of capacitance, C, is charged through a resistance R_c, from a direct current source of potential V_0 until the potential of the condenser reaches the breakdown voltage V_c of the gap between the tool and the workpiece.

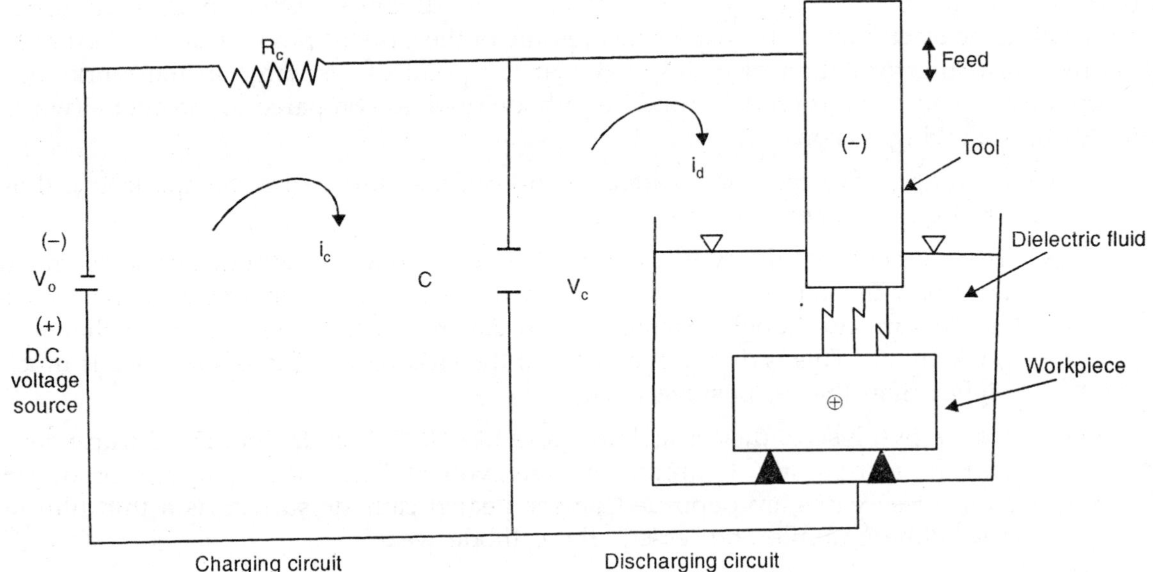

Fig. 9.7

The gap is filled with dielectric fluid. A discharge across the working gap will occur if the condenser voltage equals the breakdown voltage of the dielectric within the gap. After the discharge, the dielectric deionizes, the capacitor is recharged and the cycle repeats itself. The time taken to recharge the capacitor to the breakdown voltage must be sufficient to allow the dielectric to deionize.

The characteristics of voltage and current variation in this circuit have been shown in the figures 9.8 and 9.9.

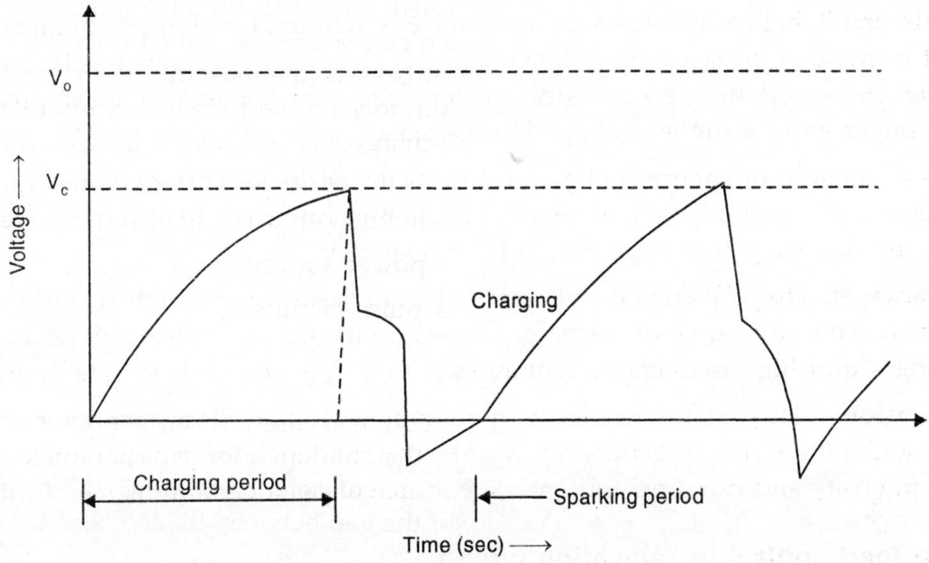

Fig. 9.8

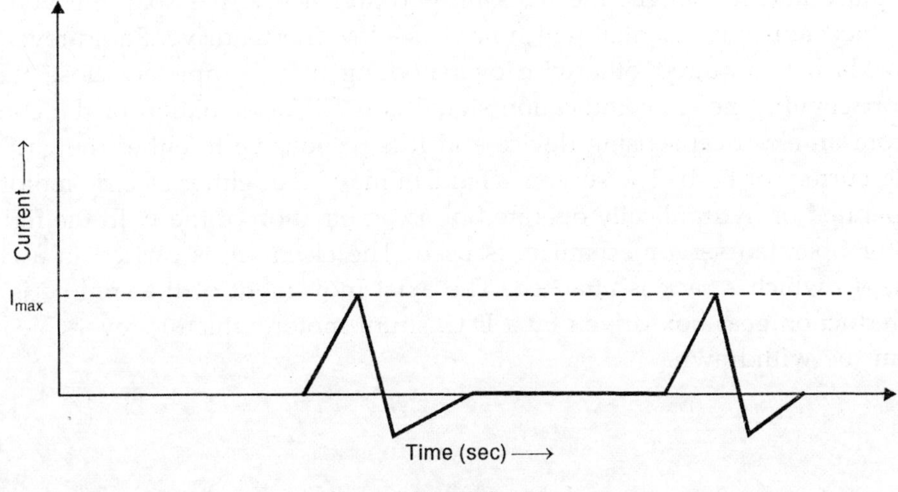

Fig. 9.9

The current $I_{max.}$ during charging depends, amongst other factors, on the capacitance of the condenser (C) and the breakdown voltage of the dielectric (V_c) and can be regulated by the spark gas size. The sparking will continue until the discharge potential of the condenser comes down below the value that is necessary for maintaining sparking. The recharging of the condenser starts as soon as sparking ceases and the entire cycle is repeated.

Resistance R_c controls the premature charging of the condenser before the deionization of the dielectric fluid layer. It reduces the condenser charging rate to a desired extent so as to avoid arcing.

In this circuit, high amperage and capacitance is required for high rate of metal removal instead of increase in the sparking frequency. The high energy sparks causes larger overcuts with larger chips and thus necessitates greater space for flushing of wear products. To maintain longer gaps, a higher voltage is necessary.

If lesser amount of energy is stored in the capacitor, lower current values can be used and frequency of sparking will increase causing the capacitor to discharge very often. However, the discharge frequency will still be relatively low.

Advantages. The relaxation circuits are simple in design, comparatively cheaper, robust and relatively extensive range of discharge. These are the practical means of generating low energy ranges and high frequencies required for fine finishing and delicate operations.

Limitations. This circuit results in high tool wear and slow metal removal rates, compared with other types of generators. Moreover, interdependence of parameters, such as discharge intensity and duration and energy values, creates a certain degree of inflexibility.

Electrode feed control in relaxation circuit

A continuous feeding of electrode towards workpiece is required and its movement should be controlled properly because during operation erosion of electrode and workpiece both takes place and it changes the working gap and hence, the sparking voltage. Rapid response of mechanism is essential which implies a low inertia drive. Rapid reversing speed with no backlash is required otherwise overshooting may completely close the gap and cause a short circuit. The error indication signal is used for actuation of the control drive, obtained from an electrical sensing device and it is responsive to either the gap voltage or the working current or both. The servomechanism may be of either electric motor operated, solenoid operated or hydraulically operated or a combination of these. In the figure 9.10 an electric-motor operated servomechanism is used. The electrode is carried in a chuck fixed to a spindle, to which a rack is attached. The axial movement of the spindle is controlled through a reduction gear box driven by a D.C. shunt motor, which is reversible so that the electrode can be withdrawn.

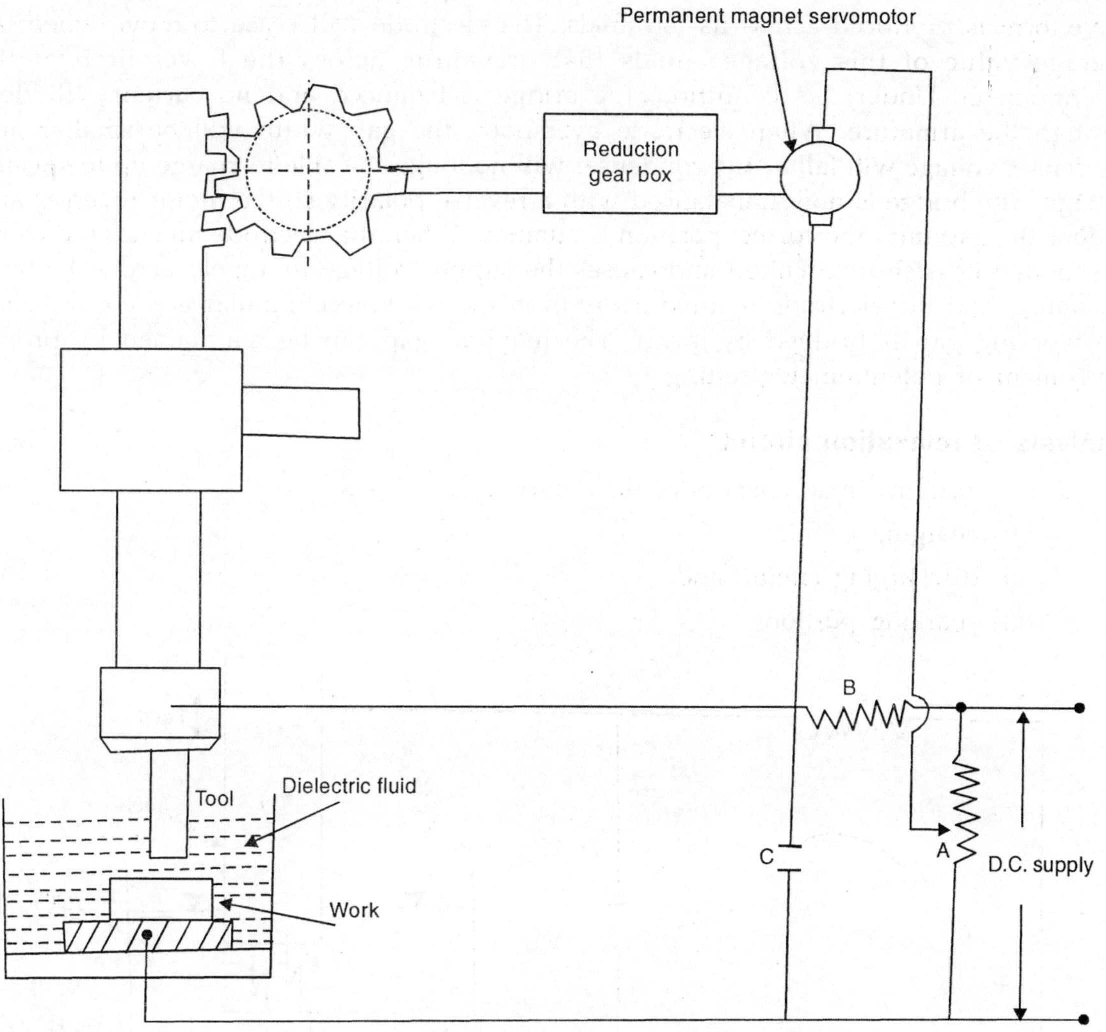

Fig. 9.10

The arms of the bridge network consist of a potential divider 'A' connected across the D.C. supply, the blast resistance 'B', condenser 'C' of the charging circuit. The motor armature is connected across a bridge network.

Initially, the electrode is widely spaced from the workpiece. When the current is switched on to the condenser, the voltage of the condenser will rise to approach the supply voltage. The supply voltage will prevail across one lower arm of the bridge while voltage across the other arm of the bridge will depend on the potentiometer setting, and if this setting is midway, then the voltage across the bridge (*i.e.* the difference between voltages across the two lower limbs) will be half the supply voltage. This voltage tends to rotate the motor, causing the electrode to close the gap. When the electrode reaches the correct position, sparking takes place and the condenser rapidly charges and discharges and a saw-tooth wave-form is produced across its terminals. The electrode will cease to move when the average value of this voltage equals that prevailing across the lower limb of the potentiometer. Under this condition, the bridge is balanced and no current will flow through the armature. When electrode overshoot, the gap width will be smaller and condenser voltage will fall as the condenser will no longer be able to charge up to specific voltage. The bridge is now unbalanced with a reverse polarity so the motor reverses and widens the gap until the correct position is attained. When the electrode touches the work, the condenser is short circuited and causes the supply voltage to appear across the blast resistance, and the electrode is lifted away from the workpiece. Similar is the case when the working gap is bridged by swarf. The required gap can be maintained by proper adjustment of potentiometer setting.

Analysis of relaxation circuit

The relaxation circuit consists of three parts :

(*i*) charging circuit,

(*ii*) discharging circuit, and

(*iii*) sparking portion.

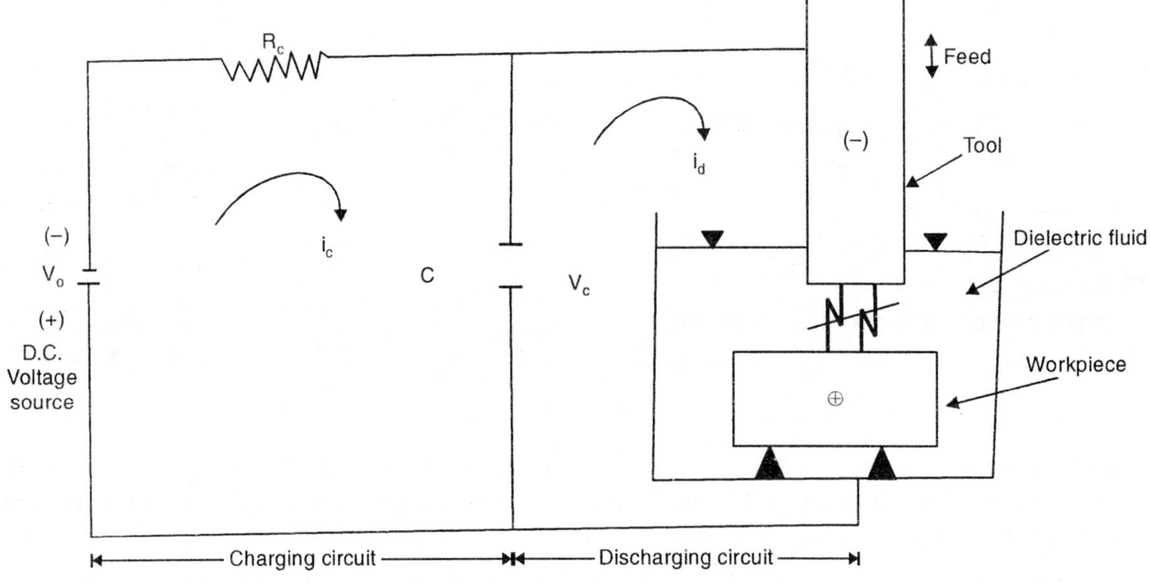

Fig. 9.11

Let

V_o = applied voltage across the circuit for charging the condenser, volts.

V_c = charged voltage of the condenser, volts.

R_c = charging resistance, ohm.

C = capacitance of the condenser, farad.

The current flowing through the charging circuit at any instant of time, t, is given by

$$i_c = \frac{V_o - V_c}{R_c} \qquad \text{...(1)}$$

Also
$$i_c = C\frac{dV_c}{dt} \qquad \text{...(2)}$$

$$\frac{V_o - V_c}{R_c} = C\frac{dV_c}{dt}$$

or,
$$\frac{dV_c}{V_o - V_c} = \frac{1}{R_c C}dt \qquad \text{...(3)}$$

On integration of equation (3) with respect to t

$$\log_e(V_o - V_c) = -\frac{t}{R_c C} + K_m \qquad \text{...(4)}$$

Where K_m is a constant.

Considering the boundary condition, at $t = 0$

$$V_c = 0$$
$$K_m = \log_e V_o$$

Thus, equation (4) becomes

$$V_c = V_o(1 - e^{-t/R_c C}) \qquad \text{...(5)}$$

The factor $R_c C$ is defined as time constant of the circuit and is equal to the time taken by a condenser to reach 0.638 times of its charging voltage, V_o.

The charging current, i_c is given by

$$i_c = C\frac{dV_c}{dt}$$

putting the value of V_c from (5)

$$i_c = C\frac{d}{dt}\left[V_o\left(1 - e^{-t/R_c C}\right)\right] \qquad \text{...(6)}$$

or,
$$i_c = \frac{CV_o e^{-t/R_c C}}{R_c C}$$

Hence,
$$i_c = \frac{V_o}{R_c}e^{-t/R_c C} \qquad \text{...(7)}$$

So, the current is a time dependent phenomenon.

Condition for maximum power delivery

The energy delivered at any period of time,

$$dE_n = i_c V_c dt$$

$$dE_n = \frac{V_o e^{-t/R_c C}}{R_c} \cdot V_o (1 - e^{-t/R_c C}) dt$$

or, $$dE_n = \frac{V_o^2}{R_c}\left[e^{-t/R_c C} - e^{-2t/R_c C}\right] dt \qquad \text{...(8)}$$

On integration, the total energy stored is given by

$$E_n = \frac{V_o^2}{R_c}\left[-\tau e^{-t/\tau} + \frac{\tau}{2} e^{-2t/\tau}\right] + K_r \qquad \text{...(9)}$$

Where, $\tau = R_c C$ and K_r is any constant.

Considering the boundary condition, at $t = 0$, $E_n = 0$, the equation (9) transforms to

$$E_n = \frac{V_o^2 \tau}{R_c}\left[\frac{1}{2} + \frac{1}{2} e^{-2t/\tau} - e^{-t/\tau}\right] \qquad \text{...(10)}$$

Suppose E_n represents the energy delivered to discharging circuit for time $t = \tau_c$, the average power delivered, $P_{d_{avg}}$ is given by

$$P_{d_{avg}} = \frac{V_o^2}{R_c}\left(\frac{\tau}{\tau_c}\right)\left[\frac{1}{2} + \frac{1}{2} e^{-2\tau_c/\tau} - e^{-\tau_c/\tau}\right] \qquad \text{...(11)}$$

For maximum power,

$$\frac{dP_{d_{avg}}}{d(\tau_c / \tau)} = 0$$

Which provides the condition that

$$\frac{\tau_c}{\tau} = 1.26 \qquad \text{...(12)}$$

Thus, $$V_c = V_o(1 - e^{-1.26})$$

Hence, $$V_c = 0.76 \text{ V}_o \qquad \text{...(13)}$$

In shows that the charging voltage will be about three-fourth of the supply voltage for transmitting the maximum power, as is evident from the figure 9.12.

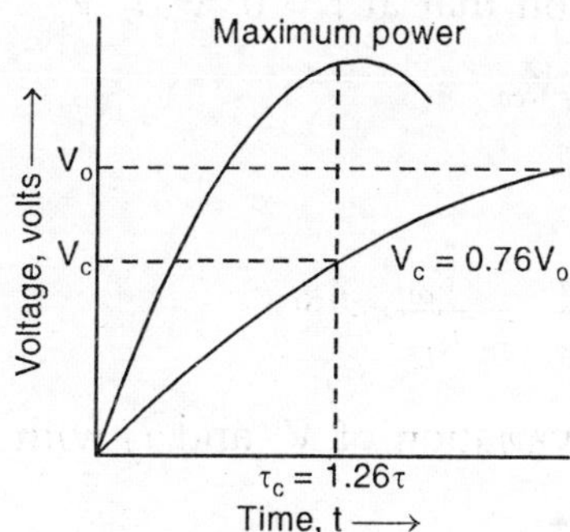

Fig. 9.12. *Condition for maximum power transmission*

Now, suppose $R_{L,S}$ represents the total resistance of the discharging portion comprising lead resistance (R_L) and sparking resistance (R_S).

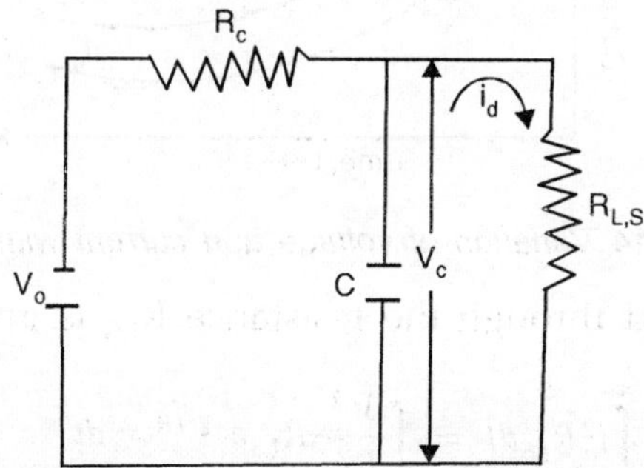

Fig. 9.13. *Relaxation circuit for analysing the discharging circuit*

The current i_d in the discharging circuit at any instant of time, t, is given by

$$i_d = -\frac{d_q}{dt} = -C\frac{dV_c}{dt} \qquad ...(14)$$

where, q is the quantity of charge in condenser.

Again, $$i_d = \frac{V_c}{R_{LS}} \qquad ...(15)$$

Hence, $$\frac{V_c}{R_{LS}} = -C\frac{dV_c}{dt}$$

$$\frac{dV_c}{V_c} = -\frac{1}{R_{LS}C}dt$$

or, $$\log_e V_c = -\frac{t}{R_{LS}C} + K_{on} \qquad ...(16)$$

where, K_{on} = any constant.

Applying boundary condition that at $t = 0$, $V_c = V_{co}$

$$K_{on} = \log_e V_{co}$$

From which, $V_c = V_{co} e^{-t/R_{LS}C}$ ...(17)

and $$i_d = \frac{V_c}{R_{LS}} = \frac{V_{co}}{R_{LS}} e^{-t/R_{LS}C}$$ ...(18)

The figure 9.14 shows the variation of V_c and i_d with respect to time.

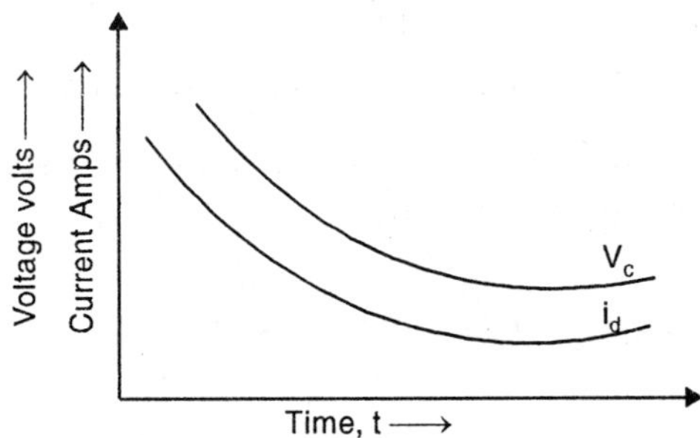

Fig. 9.14 *Variation of voltage and current with time*

Hence, energy dissipated through the resistance R_{LS} is given by

$$W_d = \int_0^{\infty} i_d^2 R_{LS} dt = \int_0^{\infty} \frac{V_{co}^2}{R_{LS}^2} R_{LS} e^{-2t/R_{LS}C} dt$$

or, $$W_d = \frac{1}{2} C V_{co}^2$$ ...(19)

R-L-C circuit

Let us consider an idealised charge-discharge circuit for an EDM process which incorporates a resistance, a capacitance and an inductance to form a R-L-C circuit.

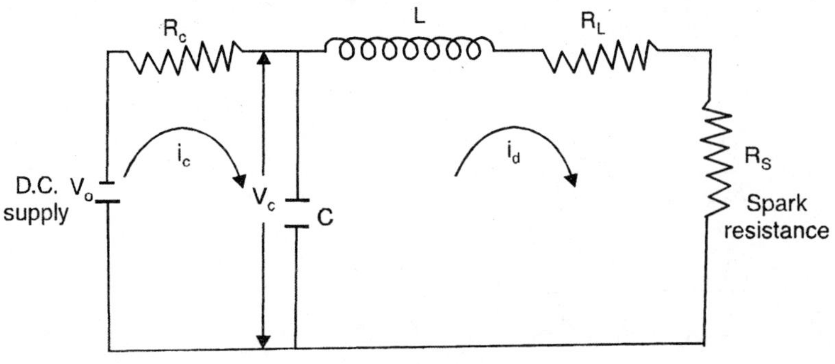

Fig. 9.15

The electrical energy supplied by the capacitor in the discharging circuit is stored in the inductance in magnetic form and hence oscillation starts.

From energy balance,

$$\frac{1}{2}CV_{co}^2 = \frac{1}{2}LI_{d\max}^2$$

or,

$$I_{d\max} = \frac{V_{co}}{\sqrt{L/C}} \qquad \text{...(20)}$$

Where $I_{d\max}$ and L are the maximum current and inductance respectively in the discharging circuit.

If e_r = self induced e.m.f due to inductance, then

$$i_d = \frac{V_c + e_r}{R_{LS}}$$

or,

$$V_c + e_r = i_d R_{LS} \qquad \text{...(21)}$$

Where R_{LS} is resultant resistance of lead and spark resistance. But

$$e_r = -L\frac{di_d}{dt} \text{ and } i_d = -C\frac{dV_c}{dt}$$

Hence,

$$V_c + LC\frac{d^2V_c}{dt^2} = -C\frac{dV_c}{dt}R_{LS}$$

or,

$$LC\frac{d^2V_c}{dt^2} + CR_{LS}\frac{dV_c}{dt} + V_c = 0 \qquad \text{...(22)}$$

If the circuit is purely inductive, then $R_{LS} = 0$,

Therefore,

$$LC\frac{d^2V_c}{dt^2} + V_c = 0$$

From which $W = \sqrt{\frac{1}{LC}}$

$$f_d = \frac{1}{2\pi}\sqrt{\frac{1}{LC}} \qquad \text{...(23)}$$

Where f_d is the discharge frequency.

From above expression, it is clear that the discharging oscillates which is undesirable because of the possibility of spark reversal. So pure inductive circuit is not desirable.

If the inductance is dominant with respect to resistance, oscillation will continue but with a decaying rate due to the presence of resistance.

Metal removal rate in relaxation circuit

From equation (5)

$$V_c = V_o(1 - e^{-t/R_cC})$$

or,
$$\frac{t}{R_cC} = -\log_e\left(1 - \frac{V_c}{V_o}\right)$$

Hence,
$$t = R_cC \log_e \frac{1}{\left(1 - \frac{V_c}{V_o}\right)} \quad \text{...(24)}$$

The frequency of charging

$$f_c = \frac{1}{t} = \frac{1}{R_cC} \frac{1}{\log \frac{1}{\left(1 - \frac{V_c}{V_o}\right)}} \quad \text{...(25)}$$

Energy delivered in each spark = $\frac{1}{2}CV_c^2$

Total energy/sec = $\frac{1}{2}f_cCV_c^2$ ...(26)

Metal removal rate in EDM, using relaxation type circuit, is proportional to the product of frequency of charging (f_c) and energy delivered per spark.

$$MRR \propto \frac{1}{2}f_cV_c^2C$$

Putting the value of f_c from (25)

$$MRR \propto \frac{1}{2}\frac{K_f}{R_c}V_c^2$$

Where
$$K_f = \frac{1}{\log_e\left[\frac{1}{1 - \left(\frac{V_c}{V_o}\right)}\right]}$$

The metal removal rate is therefore, given by

$$\text{MRR} = \frac{K_1}{2R}V_c^2\left[\frac{1}{\log_e \frac{1}{1 - \left(\frac{V_c}{V_o}\right)}}\right] \quad \text{...(27)}$$

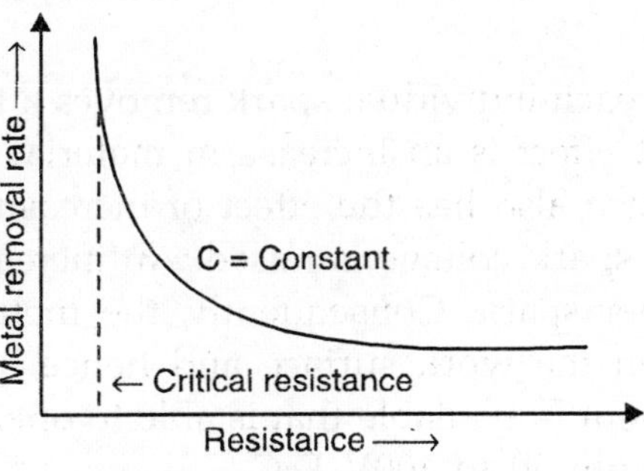

Fig. 9.16

For a given circuit, the metal removal rate will increase with decreasing R_c.

However, R cannot be made very low because, in that case, arcing will occur instead of sparking and such a situation is detrimental to the work surface finish. The minimum value of the resistance that will prevent arcing is known as critical resistance.

Concept of critical resistance

$$V_o - V_c = I_{d_{max}} R_c = \frac{V_{co}}{\sqrt{L/C}} R_c \quad \ldots(28)$$

When $I_{d_{max}}$ occurs, $V_c = 0$ for purely inductive circuit. Hence,

$$R_{c_{min}} = \frac{V_o}{V_{co}} \sqrt{\frac{L}{C}} \quad \ldots(29)$$

For maximum discharging to take place

$$\frac{V_o}{V_{co}} = 1$$

$$R_{c_{min}} = \sqrt{L/C} \quad \ldots(30)$$

The equation (30) is found to be discordant with the experimental values. In general, the circuit is not purely inductive, and hence, this equation should be used with modifications. From experimental result, it is found that

$$R_{c_{min}} \geq 30 \sqrt{L/C} \quad \ldots(31)$$

Process parameters

The above analysis of relaxation circuit shows that the factors which govern the metal removal rate are :

(*i*) Supply voltage (V_o) and breakdown voltage (V_c)

(*ii*) Charging resistance (R_c)

(*iii*) Capacitance (C)

(*iv*) Gap setting

Supply voltage

As current is increased, each individual spark removes a larger crater of metal from the workpiece. Although the net effect is an increase in material removal rates, when holding all other parameters constant it also has the effect of increasing surface roughness. Similar effect is also observed when spark voltage is raised. An increase in breakdown voltage will result in increased energy per spark. Consequently, the metal removal rate will increase, resulting in bigger craters on the work surface and hence, poor surface finish. Electrical discharge machining equipment is available that is able to operate between 0.5 and 400 amp and with voltages ranging from 40 to 400V D.C.

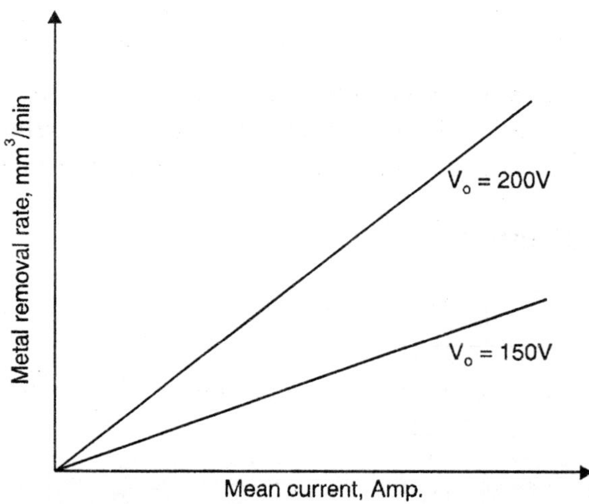

Fig. 9.17. *Effect of voltage and current on MRR*

Charging resistance

The cutting power available varies inversely as R when working gap is set constant. However, a limited value of power can be obtained by decreasing the value of R, since there will be a certain minimum value of $R(= R_{min})$ below which arcing will occur.

Capacitance

An increase in capacitance increases the energy per spark and at the same time reduces the spark frequency for a given gap setting. The cutting power available is, therefore, virtually unaltered but the surface finish deteriorates. The values of the capacitance C range between 10 and 100 microfarads.

Gap setting

The gap between the electrode and workpiece is determined by the spark voltage and current. Typical values for the gap range from 0.012 to 0.050 mm. The smaller the gap, the closer the accuracy with a better finish and slower material removal rate. If the gap is larger than the optimum, the consequent reduction in frequency is not compensated for by an increase in energy per spark and hence, the power falls. At shorter gaps the power decreases because the increased frequency is not sufficient to compensate for the reduction in stored energy. In actual practice, it is difficult to ensure optimum conditions of gap setting as power is increased because :

(*i*) The dielectric gets contaminated with metal particles and breakdown will occur at lower voltage.

(*ii*) With increasing power, R is reduced. This helps in dielectric breakdown at lower voltages.

Rotary impulse generator

In this case, power is supplied from a rotating motor generator set. The basic circuit is shown in figure 9.18. During operation, the capacitor 'C' is charged through the diode 'D' on half cycle. While the sum of the voltage supplied by the generator and the voltage of the charged capacitor is applied to the spark gap during the next half-cycle.

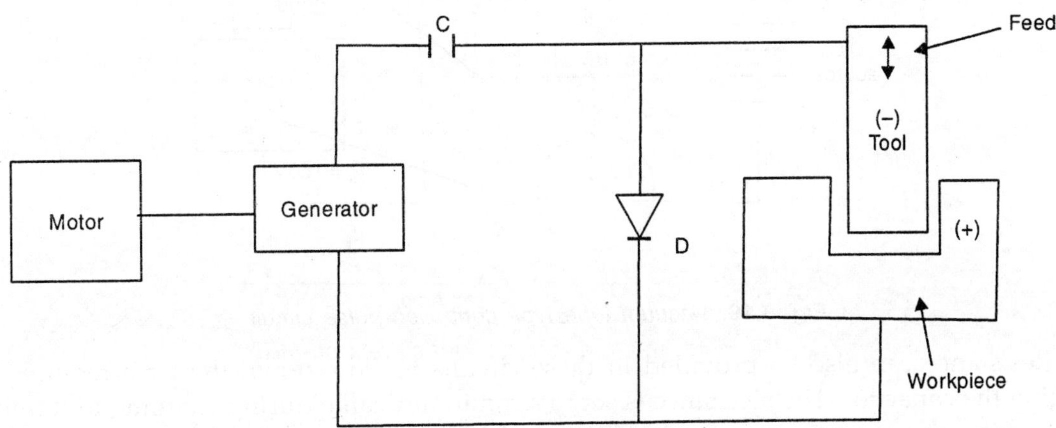

Fig. 9.18. *Rotary impulse generator*

In some applications, a non-symmetrical waveform can be generated such that the voltage in those cases will be only sufficient to cause the breakdown of the spark gap in one direction.

In other cases, unidirectional current pulses can be produced in the spark gap using sinewave generator with several combinations of semi-conductor diodes.

In rotary impulse generator, high metal removal rates is achieved but surface roughness will be poor owing to low operating frequencies.

Controlled pulse circuits

In the electrical circuits mentioned earlier, the current flow during the occurrence of short circuit in the machining gap was stopped by withdrawing the tool from the work surface by mechanical means. But, such process of mechanical retraction of tool was not very fast and ultimately resulted in undesirable burning and melting of electrodes.

In order to avoid the unwanted flow of current quickly in the event of short circuits, the electronic circuits were developed where electronic tubes and transistors were used to achieve a pulsing switch effect. These circuits are commonly known as controlled pulse circuits.

The majority of spark erosion machines currently available employ transistorised pulse circuits which can achieve higher metal removal rates together with a high degree of accuracy. The use of controlled pulse generators enables wide variations in pulse duration frequency and in the intensity of spark discharges and employs power transistor triodes as switching devices.

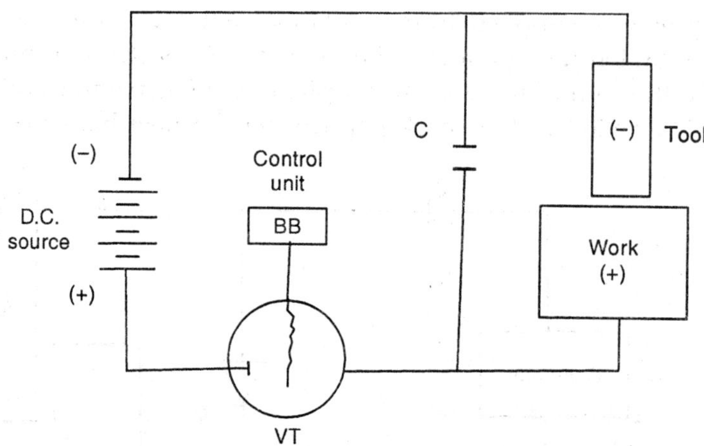

Fig. 9.19. *Vacuum tube type controlled pulse circuit*

Resistance can also be provided in these circuits to slow down the time required for charging the capacitor. The resistance is set to a minimum value during charging to diminish the charging time to a large extent and to maximum value during discharging to prevent arcing across the gap.

In vacuum tube circuit, resistance R_c of relaxation circuit is replaced by a number of vacuum tubes joined in parallel and shown as a single unit (VT) in figure 9.19. The tube bank, if effectively controlled, serves as variable resistance and the grids of the tube bank function as a switching system. An electronic control circuit (BB) operates the tubes to charge the capacitor 'C' and also makes way to prevent flow of current through the gap at the time of short-circuiting.

The vacuum tube circuits require high voltages and low currents. The recent development of power supply circuits has brought forth the adoption of transistors, which can operate at low voltages and high currents, in place of larger number of vacuum tubes for maximizing metal removal rate. The switching operation is performed by an oscillator (OSC) operating at a chosen imposed frequency without the use of any capacitor. The gap conditions control the oscillator so as to make the transistor off in case of short-circuit.

The advantage of the pulse circuits is to remove metal as fast as with R-C circuitry but to produce a better surface finish. The metal removed per spark is less but the number of sparks produced per second is large.

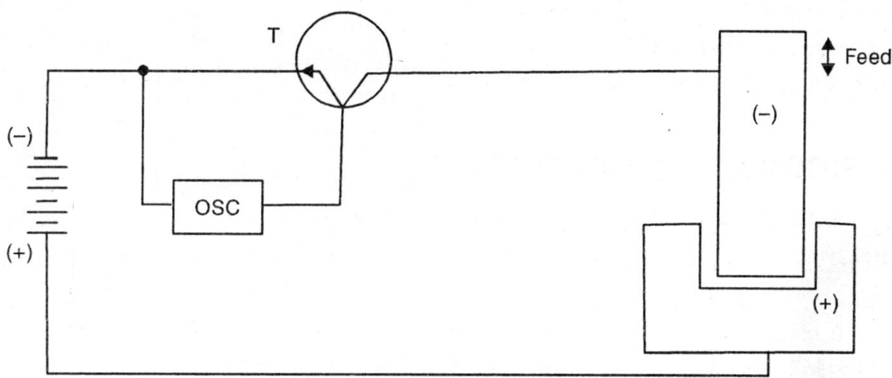

Fig. 9.20. *Pulse circuit using transistor and oscillator*

9.4 CHOICE OF DIELECTRIC FLUID

The dielectric fluid used in the EDM process :

(*i*) To act as coolant between the tool and the workpiece and to prevent arcing.

(*ii*) To serve as a conducting medium on ionisation and conveys spark.

(*iii*) To act as a coolant in quenching the spark.

(*iv*) As a flushing medium in removing chips.

The metal removal rate and electrode wear are also influenced by the dielectric fluid.

The characteristics of the dielectric fluid should be such that it gets deionize very quickly after discharging and acts as an effective insulating medium during the next charging operation of the condenser.

The dielectric fluid must fulfil the following requirements :

(*i*) It should remain electrically non-conductive until the required breakdown voltage is reached, that is, they should have high dielectric strength.

(*ii*) Breakdown electrically in the shortest possible time once the breakdown voltage has been reached.

(*iii*) Quench the spark rapidly or deionize the spark gap after the discharge has occurred.

(*iv*) Provide an effective cooling medium.

(*v*) Be capable of carrying away the swarf particles in suspension, away from the working gap.

(*vi*) Have a good degree of fluidity.

(*vii*) Be cheap and easily available.

(*viii*) It must be non-inflammable, inexpensive and chemically inert with respect to tool material, workpiece material, container, work holding devices.

The dielectric fluids which are used in general are oil, white spirit, transformer oil or kerosene. In any case, the fluid must be a hydrocarbon and it has been found that deionization occurs owing to the presence of hydrogen evolved from the hydrocarbon as decomposed.

9.5 ELECTRODES FOR EDM PROCESS

The electrode in the EDM process is the tool that determines the shape of the hole or cavity generated on the workpiece. One of the important advantage of EDM over conventional process is that a softer tool material, having a good electrical conductivity, can be effectively used for machining materials of any hardness or toughness. The selection of proper tool material in EDM process is influenced by the following criteria.

(*i*) Size of electrode and volume of material to be removed.

(*ii*) Surface finish required.

(*iii*) Tolerances desired.

(*iv*) Number of cavities to be produced.

(*v*) Nature of coolant application *i.e.* whether the coolant is flushed through the workpiece or through the electrode.

From purely technical considerations, silver-tungsten alloy is the most efficient electrode providing a high metal removal rate and very high wear ratio, but the cost of such an electrode under most condition is, prohibitive. Keeping in mind technical and economic considerations various material that can be used as tool electrodes are given in table below.

TABLE 9.1. Selection of Electrode Material

Material	Wear ratio	Metal removal rate	Fabrication	Cost	Application
Coppper	Low	High on rough range	Easy	High	On all metals
Brass	High	High only on finishing range	Easy	Low	On all metals
Tungsten	Lowest	Low	Difficult	High	Only where small holes are to be drilled
Tungsten copper alloys	Low	Low	Difficult	High	Used for higher accuracy work
Cast iron	Low	Low	Easy	Low	Can be used only on few materials
Steel	High	Low	Easy	Low	Can be used for finishing work only
Copper graphite	Low	High	Very delicate and hence difficult	High	Can be used on all metals
Zinc based alloy	High	High on rough ranges	Easily die casted	Low	Can be used on all metals

The design aspects should also be considered while selecting the electrode. The tool electrode must be designed as a mirror image of the work to be produced. However, a certain amount of clearance should be provided between the tool and work cavity produced. The magnitude of the clearance varies with the rate of metal removal, the materials of the tool and work.

9.6 FLUSHING

The dielectric fluid also acts as a flushing medium to remove the metal by-products from the cutting gap. Flushing is the most critical function for optimum process efficiency. Poor flushing results in stagnation of the dielectric fluid and a buildup of tiny machining residue particles in the gap. Stagnation usually results in low material removal rates or short circuits.

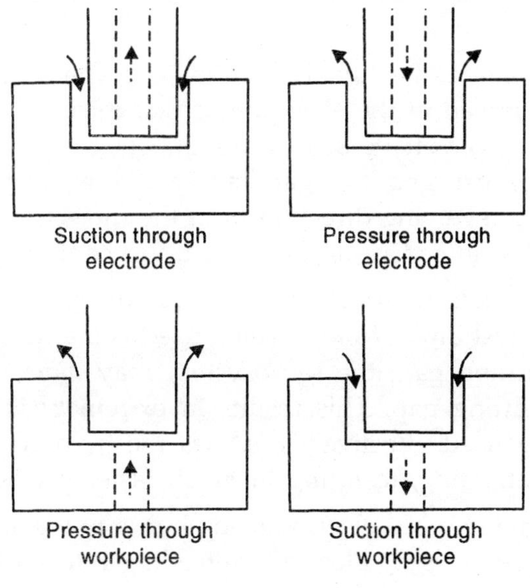

Fig. 9.21

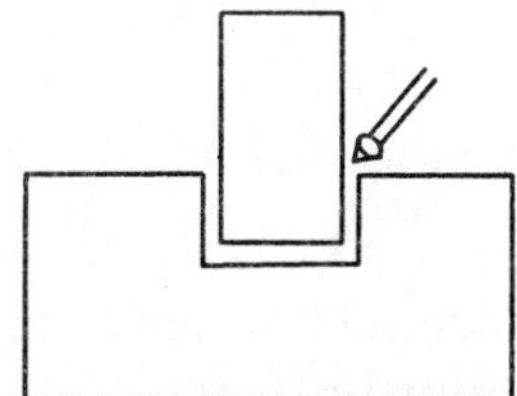

Fig. 9.22 *Jet flushing*

There are several methods available for flushing the dielectric fluid through the cutting zone. Either pressure or suction can be used with equally good results. Fluid flow through a hole down the centre of the electrode is often the easiest method, but when machining cavities and holes that do not penetrate all the way through the workpiece, a protruding "bump" will be generated in the bottom of the machined form. For example, the use of a piece of thin walled metal tubing to drill a flat-bottom hole will result in a hole with a "spike" projecting from the bottom. To avoid this problem, solid electrodes may be used or jet flushing may be incorporated.

The use of nozzles directed at the cutting gap (jet flushing) is the least desirable flushing method because of a reduction in flushing effectiveness and a corresponding reduction in material removal rates. Jet flushing should only be used if all other flushing methods are precluded by electrode or workpiece configurations. The various flushing techniques can be performed as shown in figures 9.21 and 9.22 or with the workpiece submerged in a tank of dielectric fluid. Whenever flammable dielectric liquids are being used, submersion of the workpiece is recommended to reduce the chances of accidental dielectric fluid fires.

9.7 SERVOSYSTEM

The servosystem is commanded by signals from the gap voltage sensor system in the power supply and controls the infeed of the electrode or workpiece to precisely match the rate of material removal. If the gap voltage sensor system determines that a piece of electrically conductive material has bridged the gap between the electrode and workpiece, the servosystem will react by reversing direction until the dielectric fluid flushes the gap clear. When the gap is clear, the infeed resumes and cutting continues.

The selection of the dielectric-flushing technique has a direct effect on the function of the servosystems. If the flushing technique being used is inefficient in removing the process by-products from the cutting gap, the servosystem may have to spend most of the time reversing to clear the cutting gap. This results in extremely long cycles. However if the flushing technique being used effectively removes the by-products, the servosystems will spend almost no time retracting, resulting in much faster cycles.

For compensating the reduced electrode length due to wear while hole is being drilled, electrode refeed technique is used. Electrode refeeding refers to the action of moving the wire electrode out to a fixed reference point or starting position after drilling each hole. Refeeding the electrode out to this reference position can be accomplished either manually or automatically.

9.8 APPLICATIONS

Because of the ability of EDM to machine hardened materials without producing distortion, to machine fragile workpieces without breakage, and to drill holes of various shapes into curved surfaces at steep angles without drill "Wander", EDM is the process most often selected by industries confronted with challenging materials or design problems.

The approximate growth rate for EDM equipment use by various industries are tool and die shops, 50%; captive industry toolrooms, 30%; aerospace, 10%; automotive production, 5% and others, 5%.

EDM process provides an economic advantage for making stamping tools, wire drawing and extrusion dies, header dies, forging dies, intricate mould cavities etc. It has been extensively used for machining of exotic materials used in aero-space industries, refractory metals, hard carbides, and hardenable steels.

Delicate workpieces, such as copper parts for fitting into vacuum tubes can be produced by this method. Sometimes accuracy requirements dictate the use of EDM for two main reasons :

(*i*) When repetitive shapes are required, they can often be produced from an easy to make male electrode.

(*ii*) When machining accuracy must be maintained after heat-treatment of the part.

9.9 ELECTRICAL DISCHARGE WIRE CUTTING

Introduction

Electrical discharge wire cutting is commonly known as wire-EDM which is used to produce complex two and three dimensional shapes through electrically conductive workpieces.

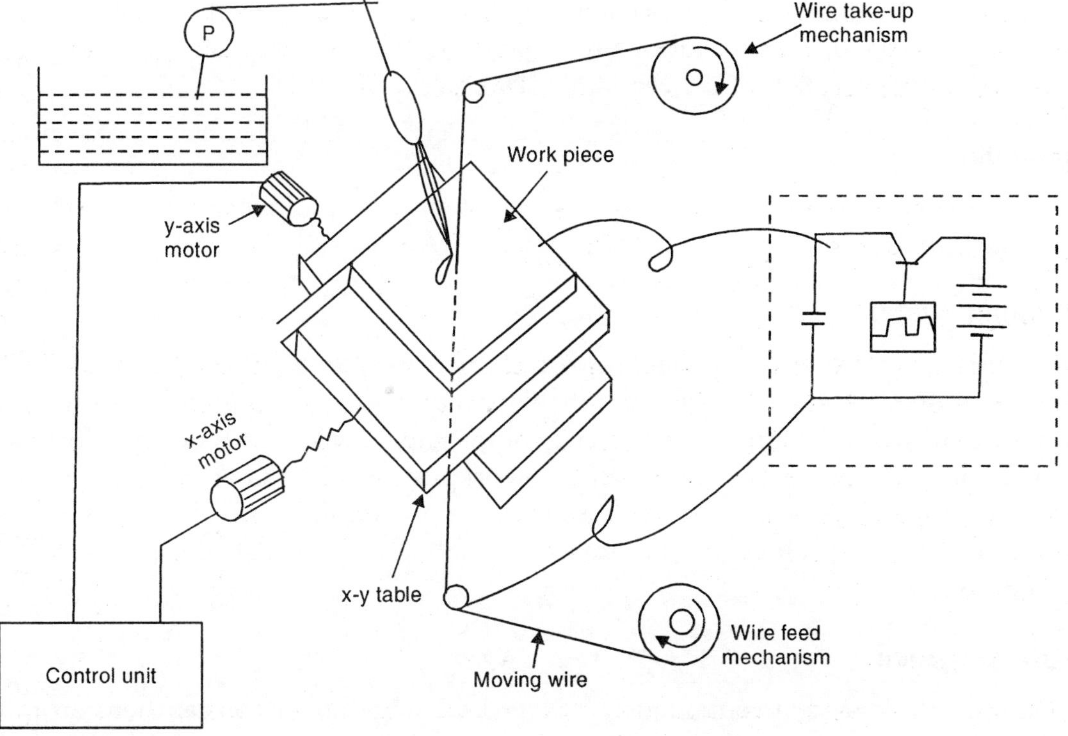

Fig. 9.23. *CNC machining with wire EDM*

Wire-EDM, comparatively new in manufacturing was first used in 1968. By 1975, its popularity was rapidly increasing because the process and its capabilities were becoming better understood by industry.

Mechanism

Wire-EDM differs from conventional EDM in that a thin, 0.05-0.30 mm in diameter wire performs as the electrode.

The wire unwinds from a spool, feeds through the workpiece, and is taken up on a second spool. A D.C. power of high-frequency pulses are supplied to the wire and the workpiece. The gap between the wire and workpiece is flooded with deionised water, which acts as the dielectric. Unlike conventional EDM, the workpiece in wire-EDM is almost never submerged in dielectric fluid. Instead, a localized stream is used.

Metal is removed ahead of the travelling wire by spark discharges, which are identical with those in conventional EDM. Either the workpiece or the wire can be moved to cause the wire to cut similarly to a bandsaw. Mechanically there is no contact between the wire and the workpiece in wire-EDM. The wire-workpiece gap usually ranges from 0.025 to 0.05 mm and constantly maintained by a computer-controlled positioning system. A complicated shape can be made on the difficult-to-machine metals without expensive formed EDM electrodes. The high degree of accuracy and fine surface obtained by wire-EDM make this process valuable for manufacturing press stamping dies, extrusion dies, prototype parts and even for the fabrication of conventional EDM electrodes.

Equipments

A wire-EDM machine consists of four subsystems; the positioning system, the wire drive system, the power supply, and the dielectric system.

Positioning system

It consist of a CNC two-axis table and in some cases, an additional multiaxis wire-positioning system. The unique feature of this system is that it operate in an adaptive control mode to always insure the consistency of the gap between the wire and workpiece. If the wire should come in contact with the workpiece, or if a small piece of material bridges the gap and causes a short circuit, the positioning system must sense this condition and back up along the programmed path to re-establish the proper cutting gap conditions.

Wire drive system

The wire drive system continuously deliver fresh wire under constant tension to the work area. The importance of the constant wire tension is to avoid problems like taper, machining streaks, wire breaks, and vibration marks. The wire delivery systems incorporate several stages of preparation to ensure wire straightness. After supply spool, the wire passes through several wire-feed and wire-removal capstan rollers. These act to buffer the eroding zone from any disturbing influences created by the wire supply.

As the wire passes through the workpiece, it is guided by a set of sapphire or diamond wire guides. Before being collected by the take-up spool, it passes through a series of tensioning rollers.

The most popular wire materials are chosen partly by the diameter of the wire being used. When wire diameters are relatively large, *i.e.* 0.15–0.30 mm, copper or brass is used. However when very fine wire 0.03–0.15 mm in diameter, is required, the additional strength derived from molybdenum steel wire is necessary.

A major portion of the spark discharges occur at the leading surface of the wire which passes through the workpiece. As a result, the wire is no longer round after one pass through the workpiece and hence it is discarded.

Power supply

The major differences between the power supplies used for wire-EDM and conventional EDM are the frequency of the pulses used and the current. To produce the smoothest surface finish possible, pulse frequencies as high as 1 MHz may be used with wire-EDM. Such a high frequency ensures that each spark removes as little material as possible, thus reducing the size of the EDM "Craters".

The wire used in this process has small diameter and hence its current carrying capability is limited. That is why, wire-EDM power supplies are built to deliver the current less than 20 Amp.

Dielectric system

In the wire-EDM process, deionized water is used as dielectric fluid. It is having following characteristics which make it preferable to others.

(*i*) Low viscosity

(*ii*) High cooling rate

(*iii*) High material removal rate

(*iv*) No fire hazard

The low-viscosity dielectric fluid is used to ensure adequate flushing in small cutting gap used with wire-EDM. Water meets this criterion. Water can also remove heat from the cutting area much more efficiently than conventional dielectric oils.

Very high specific material removal rates can be achieved when using water as the dielectric; however the wear rate on the tool (wire) is also high.

With conventional EDM, the use of flammable dielectric oils present a fire hazard, cancelling this option. When using water for the dielectric, the fire hazard problem is eliminated.

Process parameters

The linear-cutting rate for wire-EDM is approximately 38–115 mm/hr in 25 mm thick steel or approximately 20 mm/hr in 76 mm steel. The linear speed is dependent upon the thickness of the material not upon the shape of the cut. The linear-cutting rate is the same whether a straight cut or complex curves are being generated.

The speed of the wire passing through the workpiece can vary from 8–42 mm/sec depending upon cutting conditions.

Special precautions must be taken when cutting sintered carbides and various cutting tool insert materials. These materials must be machined with as little voltage as possible to avoid electrolysis from selectively removing the cobalt binder at the cut edge.

Applications

The wire-EDM process has the ability to do the jobs that are traditionally performed by an increasingly smaller number of highly skilled die makers and without the associated high labour costs. Ability of the machine to operate unattended for hours or even days further increases the attractiveness of the process.

This process can machine material as thick as 200 mm and use the computer to accurately scale the size of the part. This ability make this process valuable for the fabrication of dies of various types.

The machining of press-stamping dies is simplified because the punch, die, punch plate and stripper all can be machined from a common CNC program. Clearances are provided by modifying the basic program with an automatic scaling routine. Because clearances can be controlled so accurately, it is possible to increase die life by as much as seven to ten times.

Without wire-EDM, the fabrication process requires many hours of electrode fabrication for the conventional EDM technique, as well as many hours of manual grinding and polishing. With wire-EDM the overall fabrication time is reduced by 37%. Another popular application for wire-EDM is the machining of extrusion dies and dies for powder metal compaction. Powder metal dies, which are typically two to four times thicker than normal dies, require the machining of details with high aspect ratios. With wire-EDM, this is easily achieved without taper and without long manual machining times.

With wire-EDM, the fabrication of conventional EDM electrodes is simplified because both roughing and finishing electrodes can be made from the same basic program, again by using the scaling features. Other common applications are the fabrication of grinding wheel form tools, profile gauges and templates.

Advantages

(*i*) No electrode fabrication required

(*ii*) No cutting forces

(*iii*) Unmanned machining

(*iv*) Cuts hardened material

(*v*) Die costs reduced by 30–70%

Disadvantages

(*i*) High capital cost

(*ii*) Recast layer

(*iii*) Slow cutting rates

(*iv*) Not applicable to very large workpiece

REVIEW QUESTIONS

1. In an EDM operation employing relaxation circuit, discuss the effects of : (*i*) charging resistance, (*ii*) gap setting, and (*iii*) capacitance on the rate of metal removal.
2. It is required to drill a hole of 5 mm diameter in a 5 mm HSS sheet incorporating relaxation circuit. The surface finish required is to be 20 micron. Determine the source voltage to be set for a condenser and resistance setting of 30 μF and 50 ohm respectively.

   Find out also the time required for drilling.
3. A square hole of 4 mm side is to be drilled in a carbide piece of 3 mm thick employing R-C circuity. Evaluate the resistance to be set for a source voltage of 150 volts d.c. and a condenser setting of 50 μF. Find out also the surface finish and taper obtained out of the process. Determine the frequency of the operation.
4. How does EDM machine employing relaxation circuit compare with the pulse generator ?
5. In an EDM with R-C circuit, the supply voltage is 100 V and the break-down voltage corresponds to maximum power delivery conditions. If the supply voltage is increased to 125 V without altering any of the other electrical parameters of the circuit, what is the percentage increase in metal removal rate that can be expected ?
6. Discuss the factors influencing the choice of electrode material in EDM. Name the best electrode material for finish machining a small die made of WC.
7. Why is Flushing important in Electric Discharge Machining (EDM) ? Give a schematic diagram or EDM process and explain its limitations.
8. Assuming that, in an EDM process employing R-C type of pube generator, the gap break-down occurs at about 3/4th of the source voltage. What will be the sparking frequency, if the source voltage = 60 V, R = 3 Ω and C = 10 μF. Derive the formula used.

9. Explain the difference between the working of R.C. and R.L.C. types of pube generators used in EDM. List some of the advantages of RLC generators.
10. Explain the mechanism of metal removal in EDM process.
11. Why is Flushing important in Electric Discharge Machining ? Explain and discuss how it is accomplished.

Electron Beam Welding

10.0 INTRODUCTION

Electron beams have many special characteristics which make them most suited for specific applications. The most important of these characteristics is the high resolution and the long depth of field that is obtained because of the short wavelength of high energy electrons. Other features of the beams include their extraordinary energy (*e.g.* power densities of 10^6 KW/cm^2 have been achieved), ability to catalyse many chemical reactions, controllability and compatibility with high vacuum.

Electron beam welding is a high energy density fusion welding process that utilizes a tightly focused, high-velocity stream of electrons as the thermal source.

The extensive use of electrons in this process is due to the fact that electron, the smallest, stable, charged elementary particle of matter can be very easily obtained in the free state. The surface of a metal can emit electrons when sufficient energy is supplied. In order to obtain free electrons, thermo-electronic cathodes are used wherein metals are heated to the temperature at which the electrons acquire sufficient speed for escaping to the space around the cathode.

In the late 1800s, W.C. Roentgen found that when a beam of electrons were suddenly stopped by impact with a target, a secondary emission of X-rays was generated. He also found that the electron beam heated and melted the target.

In 1940s. individuals in the manufacturing community began to consider the possibility of using electron beams to perform various fabrication tasks. In 1952, the first such electron beam machine for use as a welding tool was built at Carl Zeiss Gmbh by Dr. H.C. Karl Heinz Steigerwald.

10.1 BACKGROUND OF ELECTRON BEAM ACTION

Free electrons under the effect of electric or magnetic fields may be accelerated greatly. Even with relatively small potential difference, electron velocity can be of the same order as the speed of light.

Let an electron, moves in an electric and magnetic field is subjected to Lorenz force of

$$F_L = \vec{F_e} + \vec{F_m}$$

Where, $F_e e = -eE_s$ = force from the electrostatic field of E_s

$F_m = -e\mu V_s H$ = force on the electron due to magnetic field

μ = magnetic permeability

e = charge of electron

H_s = strength of the magnetic field

V_s = speed of the electron

The magnetic force is always at right-angles to the direction of electron movement. Hence, it does no work, but can only change the direction of the electron movement. Thus the kinetic energy of the electron is not affected by the magnetic field having no component acting in the direction of the electron movement.

The sum of the kinetic and potential energies of electrons in motion in a given electric field is constant, when

$$\frac{1}{2} m_e V_s^2 - eE_s = C_o$$

At the terminal conditions, $V_s = 0$ when $E_s = 0$; from which, $C_o = 0$. Thus,

$$\frac{1}{2} m_e V_s^2 = eE_s$$

from which the speed of electron is given by

$$V_s = \sqrt{\frac{2e}{m_e} E_s} \text{ m/sec.}$$

Where m_e = mass of electron, Kg

e = charge of electron, coulomb

E_s = Voltage of the electric field, volt

The static mass of electron is 9.1×10^{-28} gm and $e = 1.6 \times 10^{-19}$ coulomb.

$$\therefore \quad V_s = \sqrt{\frac{2 \times 1.6 \times 10^{-19}}{9.1 \times 10^{-28} / 10^2}} \sqrt{E_s} = 600 \sqrt{E_s}$$

Thus, the variation of the speed of electron is a function of the potential difference of the electric field. The order of this velocity can be calculated with velocity of light, when

$$\frac{V_S}{C_L} = \frac{600\sqrt{E_S}}{930 \times 10^3} = 0.000,645 \sqrt{E_s}$$

Where C_L is the speed of light in Km/Sec.

If the applied voltage lies in the range of 10^6 to 10^8 volts, the order of electron velocity would lie in the range of the velocity of light thus imparting a very high amount of kinetic energy to the electrons.

Thus, electrons moving in an electric field can accumulate considerable kinetic energy. When the electrons reach the anode surface, a rapid change in momentum takes place through collisions with the atoms of the workpiece metal when the kinetic energy of the electrons is transferred to the atoms of the work material under bombardment. The various technological processes like micro-welding, melting and cutting and making thin films can be carried out by controlling the concentrated thermal energy obtained through electron bombardment.

10.2 PRINCIPLES OF ELECTRON BEAM WELDING

The electron beam welding process takes place when a high-velocity beam of electrons is directed at a workpiece. As the electrons impact the workpiece material, their kinetic energy is converted into thermal energy with an efficiency of nearly 100%.

The electron beam welding is generally performed in a welding chamber that has been pumped to a high vacuum. This is necessary to ensure optimum beam propagation and focussing. Because electrons have mass, they interact with air molecules, which results in beam dispersion and a large loss of energy. A secondary benefit of performing the EBW process in a vacuum environment is that the molten material in the weld nugget undergoes a vacuum melting and refining process. This results in a high weld material purity and a favourable weld nugget microstructure.

Electron beams are focused by means of magnetic fields to exceedingly small sizes and hence it is called a high energy density welding process. This welding process is able to concentrate their thermal energy into small area and is capable of high welding rate which, in turn, result in a very low levels of workpiece distortion. Additionally, if sufficient power is available in the heat source, exceptionally thick section of material can be welded in a single pass.

The welding processes with low power density such as TIG welding are very limited in the single pass penetration that can be achieved. This limitation is influenced largely by the thermal conductivity of the material being joined. Welding processes using power densities less than $10^5 W/cm^2$ are operating in the conduction-welding mode. Therefore when thick sections of material are to be joined by a conduction-welding process, the edges of the material must be prepared by bevelling and then the resulting gap must be filled with many individual weld passes.

When the electron beam impinges on the workpiece, a thin layer of material is instantaneously melted. If sufficient power density is available, a small amount of this molten material will be vaporised, thus displacing a portion of the molten material to the sides of the beam. As the molten material is displaced, material that is still solid becomes exposed to the rapid melting action of the beam. Thus the electron beam drills a small hole that penetrates deep into the workpiece.

The fusion zone for an electron-beam welded joint is only 5–10% of that produced when performed by arc-welding processes. This reduction in the size of the fusion zone leads to a corresponding decrease in the total heat input and distortion of the part being welded, alongwith a decrease in the degradation of the properties of heat-sensitive materials and an increase in the weldability of high thermal conductivity materials.

10.3 GENERATION AND CONTROL OF ELECTRON BEAM

There are three major components of an electron beams welding system. These components work together to generate the electron beam and to provide the optimum environment for the process.

(*i*) Electron beam gun

(*ii*) Power supply

(*iii*) Vacuum system.

Electron beam gun

The electron gun emitting electron beam consists of

(*a*) A cathode, which serves as the source of electrons. It may be a current-carrying, self heating filament, or a solid block indirectly heated by radiation from a filament.

(*b*) A grid cup, which is negatively biased with respect to the filament.

(*c*) An anode which is kept at ground potential, and through which the high velocity electrons pass.

The cathode or filament, which provides the source of electron emission is constructed of a shaped strip of tungsten or tantalum and is heated to a temperature of approximately 2480°C. The filament at this temperature causes a thermionic emission of electrons. The filament, acting as a hot cathode emitter, is maintained at a high negative potential to repel the negatively charged electrons toward the ground potential anode.

As the newly formed beam of electrons begins its acceleration to the anode, it passes through a hole in the bias grid. A negative-biasing voltage applied to the grid controls the flow of electrons to the anode. By varying the negative potential difference between the grid and the cathode, the flow of electrons in the beam can be completely suppressed or precisely ramped to various discreet levels.

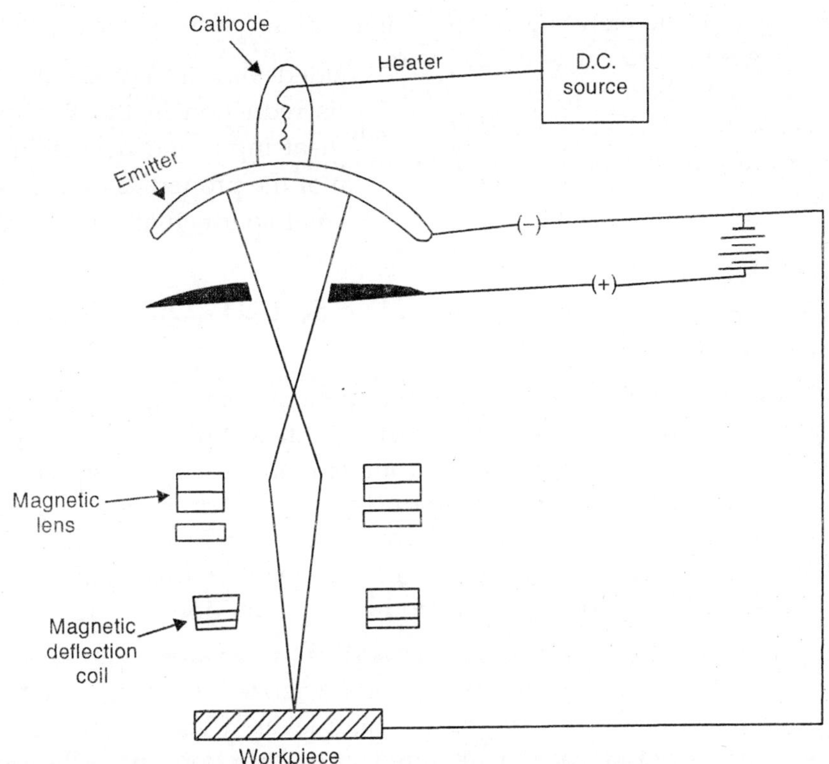

Fig. 10.1. *Schematic set-up of EBW*

The beam continues accelerating while passing through the grid until the maximum velocity is reached as it passes through a hole in the anode. The high velocity electron beam then passes through an electromagnetic focussing lens that forces the beam to converge to a focal point at the workpiece. The final set of electromagnetic coil known as the beam deflection system is used to provide a small amount of programmable beam motion to produce porosity reduction, widening of the weld zone, and seam tracking.

A high vacuum of between 1.3×10^{-2} Pa and 1.3×10^{-4} Pa is maintained in the gun assembly to provide the proper environment for beam propagation, to avoid arcing between high-voltage elements, and to reduce oxidation and increase the lifetime of the consumable high temperature filament.

Power supply

The electrical energy is supplied for heating the filament, accelerating the electrons, controlling the bias grid, focusing the beam and operating the deflection coil.

The electron beam systems, commercially available can generate accelerating voltage from 30 to 300 Kv and beam current from 0.5 μ amp to 1.5 amp. The product of these two numbers determines the total beam power, which generally ranges from 30W to 100KW. Nowadays even higher powered units are available.

Electron beam systems are typically classified as either low-voltage or high-voltage systems, referring to the accelerating voltage. High-voltage systems are capable of generating accelerating voltages between 60 and 200 Kv, rendering them capable of the deepest welding penetrations and the highest nugget aspect ratios. Low-voltage EBW systems typically range between 15 and 30 KV of accelerating voltage. Although they are limited in penetration and the ability to produce high aspect ratio welds, the lower voltage results in a reduction of the equipment cost.

Vacuum systems

Vacuum system in EBW process comprises of two subsystems : (*i*) the pumping system (*ii*) the vacuum chamber.

All EBW systems require some form of pumping system to generate a high vacuum in the electron gun. In high-vacuum electron beam welders (EBW-HV), workpiece alongwith electron gun are kept in high-vacuum environment. In this case highest degree of weld purity and cleanliness are achieved but productivity of machine is limited because of the time required to evacuate the air from the work chamber. In non-vacuum electron beam welders, the electron gun is in high vacuum, but beam is delivered to a workpiece at atmospheric pressure, thus avoiding non productive pumpdown cycles completely.

Vacuum pumps used for electron beam welders are grouped into two categories. The first category known as roughing pumps, are characterised as being able to provide high flow rates but are not capable of attaining the high level of vacuum required for EBW-HV chambers and for the level of vacuum required in the electron gun. The roughing pumps are used for removal of large quantity of gas from a volume but high vacuum is not required. Multiple vane rotary pumps, reciprocating piston pumps or roots blowers are used as roughing pumps.

The second category of pump, known as a diffusion pump is used to create the high level of vacuum required in a EBW-HV chamber or in the electron gun. To generate this high level of vacuum as fast as possible, a roughing pump is first used to rapidly reach a partial vacuum level, then a diffusion pump is used to remove the remaining air molecules, thus producing the high vacuum.

A vacuum chamber is required for all EBW-HV and EBW-MV systems not only to provide a suitable environment to carry out the welding operation but to also act as a shield to block the harmful X-rays that are generated when the electron beam, strikes the workpiece. The higher the accelerating voltage of the electron, beam the stronger the X-rays.

The protection from X-ray is done by constructing the vacuum chamber of low-voltage EBW systems with 12.7 mm thick stainless steel or mild steel. High-voltage EBW systems require the addition of lead sheets to the chamber walls.

10.4 PROCESS PARAMETERS

The electron beam welding process is governed by the various process parameters such as :

(*i*) accelerating voltage

(*ii*) beam current

(*iii*) beam power

(*iv*) traverse rate

(*v*) beam spot size

Increasing the accelerating voltage or beam current increases the depth of penetration. The product of these two variables equals the overall beam power which, in turn, determines the amount of material melted for a given welding traverse rate. Maintaining all other variables constant, an increase in welding traverse rate, decreases the depth of penetration, the energy input per unit length, and the width of the weld bead.

The diameter of the electron beam spot on the workpiece is influenced by the gun design, beam current, focus current and stand-off distance. Most electron beam systems are capable of focussing the beam to diameters as narrow as 0.25 mm; however, fit up and beam alignment become very critical with such a narrow beam. Larger spot sizes, upto 1.27 mm are used when a shallow, wide weld nugget is desired.

The level of vacuum in which welding is performed also affect the performance of the process. As pressure increases, welding penetration decreases and the width of the fusion zone increases.

10.5 PROCESS CAPABILITIES

Electron beam welding can perform a single-pass, deep penetration welds while maintaining a very low overall thermal effect. Current penetration capabilities for EBW-HV are in excess of 300 mm, while EBW-MV is limited to approximately 15 mm and EBW-NV to 12.7 mm.

It can perform multiple-tier welding; the ability to simultaneously weld two or more layers of material that are separated from each other by spaces upto several centimeters.

While welding ferrous alloys materials, parts must be demagnetised before welding to avoid unwanted beam deflection, as the electron beam is influenced by magnetic fields.

This process has the ability to do welding with large gaps in thick workpiece. A device is used to monitor the X-ray emission from the weld zone to determine the width and location of the joint gap. A wire feeder provides extra material to fill gap.

The materials that exhibit very good weldability are stainless steels, low-carbon steels, copper and copper alloys, nickel-based alloys, aluminium alloys, and refractories such as zirconium, tantalam, titanium and niobium.

10.6 APPLICATION EXAMPLES

EBW is used in fabrication of large coupling hubs that are shaped like a disk with six large clutch fingers projecting from the face-originally, these assemblies were cast to a near net shape and then machined to the final configuration. This technique was undesirable because of the difficulty in casting the complex shape and the expensive machining required to achieve the final complex configuration. By adopting a manufacturing change to cast the product in seven pieces, machine to final shape as individual pieces and join together with electron beam welding, a substantial increase in casting yield and a reduction in machining cost were realised. The electron beam welds were made with 24 KW of beam power through 70 mm thick 0.3% carbon steel at a rate of 5 mm/sec.

The bimetal strip is produced for the electronics industry by EBW. Two continuously fed strips of 0.76 mm thick silver and copper are butt welded together to form a continuous strip of bimetal. After welding, the material is then used in a stamping operation to form high-reliability electrical contacts. The copper portion becomes the surface that will be soldered, and the silver portion becomes the electrical contact surface.

The EBW–HV process is used for two-tier welding of aircraft gas turbine engine components. Two 1.9 mm thick cylindrical sub-components, which are separated by 12.7 mm, are simultaneously joined with a circumferential weld at a speed of 12.7 mm/sec. The electron beam was focused midway between the two tiers to produce a weld bead 2-mm wide in the top tier and 1.2 mm wide in the bottom tier.

A very popular application for EBW-NV is the welding of automotive catalytic converters. Earlier general motors has been using 35 KW computer controlled EBW-NV systems to perform an edge weld on catalytic converters. Four layers of 1.27 mm thick stainless steel are fused together to form the periphery of the part. Parts are fed on a conveyor into a lead-lined room where they are automatically loaded into a fixture on a rotary carousel. The linear welding speed is 110 mm/sec with a gun/workpiece distance of 25.4 mm.

10.7 PROCESS SUMMARY

Advantages

(*i*) Low thermal input.

(*ii*) High purity : EBW-HV and EBW-MV.

(*iii*) Deep penetration.

(*iv*) Welds high-conductivity metals.

Disadvantages

(*i*) High joint preparation and tool costs.

(*ii*) Non-productive pumpdown time : EBW-HV and EBW-MV.

(*iii*) Limited penetration and stand-off distance with EBW-NV.

(*iv*) X-ray shielding required.

(*v*) Seam tracking sometimes difficult.

10.8 ELECTRON BEAM MACHINING

Introduction

Electron beam machining is invented by Dr. K.H. Steigerwald in 1952 in Germany and is capable to drill materials upto 10 mm thick at perforation rates. In Electron Beam machining, the material is removed by the process that utilizes a focused beam of high-velocity electrons to perform high speed drilling and cutting. The material heating action is obtained when high-velocity electrons strike the workpiece. Upon impact, the kinetic energy of the electrons is converted into the heat necessary for the rapid melting and vaporization of any material.

PROCESS PRINCIPLE

The electron beam machining process begins after the workpiece is placed in the work chamber and a vacuum is achieved. The creation of a hole by an electron beam occurs in four stages. First, the electron beam is focused onto the workpiece to a diameter that is slightly smaller than the final desired hole diameter. Power is adjusted so that the electron beam will generate a power density at the workpiece in excess of 10^8 W/cm^2. A power density of that magnitude is more than sufficient to instantly melt any material regardless of thermal conductivity or melting point. Drilling is accomplished through the combination of an electron beam pulse and an organic or synthetic backing material, which is applied to the exit side of the surface being drilled.

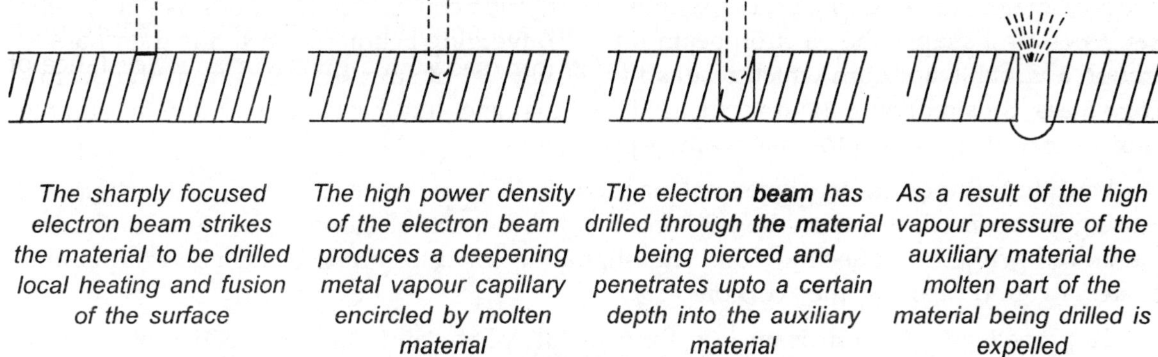

Fig. 10.2. *Four steps that lead to material removal by EB drilling*

When the focused beam strikes the workpiece, local heating, melting and vaporization take place instantly. Only about 5% of the affected material is actually vaporized. The pressure of the escaping vapour is sufficient to form and maintain a small capillary channel in the material. The beam and capillary rapidly penetrate through the workpiece and a finite distance into the backing material. The volume of backing material that is contacted by the beam is almost totally vaporised resulting in the explosive release of backing material vapour. As a result of the comparatively high pressure of the backing material vapour, the molten walls of the capillary are expelled in a shower of sparks leaving a hole in the workpiece and a small cavern in the backing material.

A single pulse is often used to produce a single hole : however if the material is very thick, multiple pulses may be required. If the desired hole shape is not round, the beam, pulsing at rates up to 1000 pulses/sec, is deflected by computer to cut out the shape along its perimeter. Using this technique, almost any hole shape can be generated.

EQUIPMENT

Most of the subassemblies of the electron beam machining are similar to that of electron beam welding such as vacuum system, workpiece positioning system, and vacuum chamber. The significant differences between the two systems are with respect to the electron beam gun and power supply.

Electron beam gun

The function of the electron beam gun is to generate, shape and deflect the electron beam to drill or machine the workpiece. The EBM gun is designed to be used exclusively for material removal applications and can be operated only in the pulsed mode.

An electron "cloud" is generated by a super heated tungsten filament, which also acts as the cathode. A combination of repelling forces from the negative cathode and the attracting forces from the positive anode causes the free electrons to be accelerated and directed toward the workpiece. Before passing through the anode, the beam travels through a bias electrode, which controls the flow of electrons and acts as a switch for generating pulses.

After passing through the anode, the electron beam is diverging rapidly and travelling at approximately one-half the speed of light. A magnetic coil, which functions as a magnetic lens, repels and shapes the electron beam into a converging beam. The beam is then passed through a variable aperture which results in the removal of stray electrons from the beam's fringe areas, thus reducing the final focused spot diameter and producing a more favourable beam energy distribution for machining applications.

Beneath the aperture are three final magnetic lens, deflection coil and stigmator. Pinpoint focusing is accomplished with the lens and a small amount of controllable beam deflection is achieved with the deflection coil. The stigmator corrects minor beam aberrations and ensures a round beam at the workpiece.

To protect the electron beam gun from metal spatter and vapour, a series of rotating slotted disks are often mounted directly beneath the gun exit opening. The rotational rate of the disks is synchronised such that the beam pulse will pass through the slots, but the spatter is blocked.

Power supply

The high-voltage power supply used for EBM systems generates voltages of upto 150 KV to accelerate the electrons. The most powerful electron beam machining systems are capable of delivering enough power to operate guns at average power levels of up to 12 KW. Individual pulse energy can reach 120 joules/pulse. To avoid the possibility of arcing and short circuits, the high-voltage sections of the power supply are submerged in an insulating dielectric oil.

All power supply variables, such as the accelerating current, focus current, pulse duration and others, are controlled by a CNC unit or by a microcomputer. To ensure process repeatability, the process variables are monitored and compared with set-points by the power supply computer. If a discrepancy arises the operator is alerted.

PROCESS PARAMETERS

There are four important parameters associated with electron beam machining namely beam current, pulse duration, lens current and the beam deflection signal. Each parameter is computer controlled during process after initial setting which is done by trial-and-error testing.

Beam current is continuously adjustable between 100 μ amp. to 1 amp. As the beam current setting is increased, the energy per pulse delivered to the workpiece is also increased. It can generate energy in excess of 120 joules/pulse.

Pulse duration affects both the depth and the diameter of the hole. The longer the pulse duration, the wider the diameter and the deeper the drilling depth capability will be. To a certain extent, the amount of recast and the depth of the heat-affected zone will be governed by the pulse duration. Typically, electron beam systems can generate pulses as short as 50 μ sec or as long as 10 m sec.

The lens current parameter determines the distance between the focal point and the electron beam gun (working distance) and also determines the size of the focused spot on the workpiece. The diameter of the focused electron beam spot on the workpiece will, in turn, determine the diameter of the hole produced. To achieve a desired hole diameter, the beam power must be sufficient to generate more than $10^8 W/cm^2$, otherwise the power density will be insufficient to promote vaporization and drilling.

The depth to which the focal point is positioned beneath the workpiece surface determines the axial shape of the drilled hole. By selecting different focal positions, the hole produced can be tapered, straight, inversely tapered, bellshaped, or center-bowed.

When hole shapes are required to be other than round, the beam deflection coil is programmed to sweep the beam in the pattern necessary to cut out the shape at the hole's periphery. Beam deflection is usually applicable only to shapes smaller than 6 mm.

PROCESS CAPABILITIES

A wide range of materials, such as stainless steel, nickel and cobalt alloys, copper, aluminium, titanium, ceramics, leather and plastics, can be successfully processed by EBM. There are some thermal effects remain on the machined edge after processing. However, because of the extremely high beam power density and the short duration of the beam/ workpiece interaction time, thermal effects are usually limited to a recast layer and the heat-affected zone, which seldom exceeds 0.025 mm. There is no burr generated on the exit side of the hole.

Electron beam process has the ability to drill deep, high aspect ratio holes due to its high power density. The hole diameters that can be drilled range from 0.1 to 1.4 mm in thickness upto 10 mm. The tolerance on the hole diameter is ± 5% of the diameter or 0.03 mm, whichever is greater.

An electron beam does not apply any force to the workpiece, thereby allowing brittle or fragile materials to be processed without danger of fracturing.

APPLICATION EXAMPLES

EBM process is suitable where thousands of simple holes to be drilled in each workpiece, or those that require more than 200 holes that are difficult to drill conventionally because of material hardness or hole geometry. Most current applications of EBM are for the aerospace, insulation, food and chemical and clothing industries.

The drilling of a turbine engine combustor dome made of a Cr Ni Co MoW steel has been performed for several years using EBM.

The insulation industry relies on EBM to drill thousands of small holes in cobalt alloy fibre spinning heads that are used in the production of glass fiber and rockwool materials.

Filters and screens used in the food-processing industry require thousands of holes to be drilled through relatively thin formed sheet metal.

REVIEW QUESTIONS

1. Write in brief the summary of Electron Beam Machining (EBM) process.
2. What are the different equipments used in Electron Beam Welding (EBW) ? Explain with neat sketches.
3. Write four specific applications where you feel that EBM should be the preferable choice. Explain the effect of "focussing" on the performance of LBM.

Laser Beam Processing

11.0 INTRODUCTION

Laser material processing utilizes the energy from coherent beams of light to remove, melt or thermally modify materials. Laser has a broad range of tasks that can be performed for a wide variety of industries. Apart from cutting, drilling and welding, it can also be used to mark parts, heat-treat surfaces and selectively clad materials. Laser is having very high processing speeds, low thermal effect on the workpiece, and its applicability to automation.

This all are possible only due to the capability of laser to produce high energy densities. If sun rays are focussed by lens system, it can burn a bundle of paper, the energy density being of the order of 100 watt/cm^2. A laser beam can melt and vaporize diamond when focussed by lens system, the energy density being of the order of 100000 KW/cm^2. To create such intensity from sun, the sun would have to change from its normal temperature of 6000°K to about 100000°K.

Such tremendous energy release is due to certain atoms which have higher energy level and oscillate with particular frequency. When such atoms impinge with electro-magnetic waves having resonant frequency, the waves absorb energy from the atoms and become highly powerful. Such waves with increased energy are called 'Maser' (Microwave Amplification by Stimulated Emission of Radiation).

This was invented in 1917 by Albert Einstein when he hypothesized that under the proper conditions light energy of a particular frequency could be used to stimulate the electrons in an atom to emit additional light with exactly the same characteristics as the original stimulating light source.

The specialities of laser rays that led to such high capabilities and distinguish it from other light sources include :

(*i*) Spectral purity

(*ii*) Highly directive property

(*iii*) High focussed density

The light beam emitted by the laser is highly collimated with optical divergence angles limited to 10^{-2} to 10^{-4} radians. This also leads to high power density.

11.1 LASING PROCESS

The operation of a laser depends on the utilization and to some extent, manipulation of the naturally occurring transitions between the energy levels of a quantized system such as atom.

When electron is free *i.e.,* when its energy is positive, it can have any energy moving at any speed. But bound energies are not arbitrary and are discrete set of allowed value. Let these allowed values of energy level be denoted by E_0, E_1, E_2, ..., etc. If an atom is initially in any of the excited states, it does not remain in that state. Sooner or later it drops to the lower energy state and radiates energy in the form of light. The frequency of emitted light is determined from the principle of conservation of energy between the difference of energy levels and quantum–mechanical understanding of light energy, when the transition from energy E_3 to energy E_0 is related by,

$$hW_{30} = E_3 - E_0$$

Where, h = plank's constant

W_{30} = frequency of the liberated light

E_3, E_0 = energy levels between which transition occurs.

Einstein brought in a new facet in this area, called "Interaction of Radiation with matter". Considering any two discrete energy levels of an atom, pth and qth level as shown below.

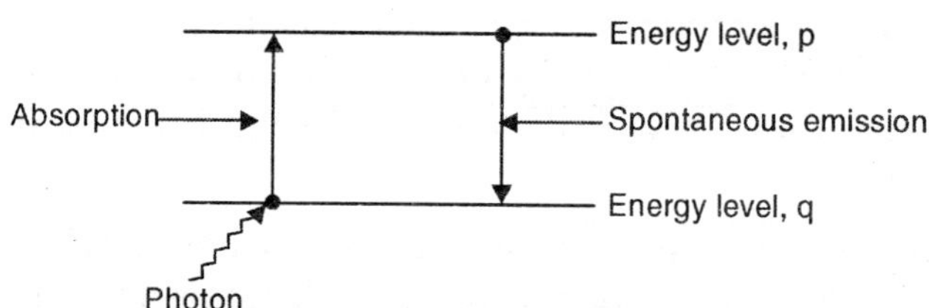

Fig. 11.1

Einstein proposed that when such an atom has light of right frequency acting on it, it can absorb that photon of light and can make a transition from state q to state p. The probability that this phenomenon occurs per second depends upon :

(*a*) the energy levels of the two states between which transition takes place, and

(*b*) the intensity of the light acting on to the atom.

The phenomenon of upwardly transition is known as "Absorption". Consider a certain number of atoms N_q in the energy level q and N_p in the energy level p are in equilibrium at temperature θ. Now, the number of atoms that are going from q to p per second by the process of absorption is given by

$$R_{q \to p} = N_q B_{qp} \, I(w)$$

$I(w)$ is the intensity of distribution of light and B_{qp} is a proportionality constant depending on the pair of energy levels between which the process is occurring. Some levels are easy to excite while other levels are hard to excite.

The next process is the case of emission when transition of an atom takes place from the energy level p to the energy level q. Einstein proposed that there are two processes by which such emission can occur.

(*i*) Spontaneous emission in which no light is present at the higher energy level and the atom in an excited state could fall to a lower state emitting a photon and the process is independent of light intensity. Thus, the number of atoms going from p to q by the process of spontaneous emission,

$$R_{p \to q/sp.\ \text{emission}} = A_{pq}N_p$$

where A_{pq} is the probability constant per unit time for spontaneous emission.

(*ii*) Stimulated or induced emission in which the emission is influenced by the presence of light. If light of the right frequency acts on an atom, an increased rate of photon will be emitted and such emissions would be proportional to the intensity of light.

Thus, if an atom or molecule is raised to a high energy level p from lower energy level q by an outside energy source (*e.g.*, heat, light, chemical reaction, etc.), it begins to radiate this acquired energy spontaneously emitting a photon while reverting back to the ground state or any other lower energy level.

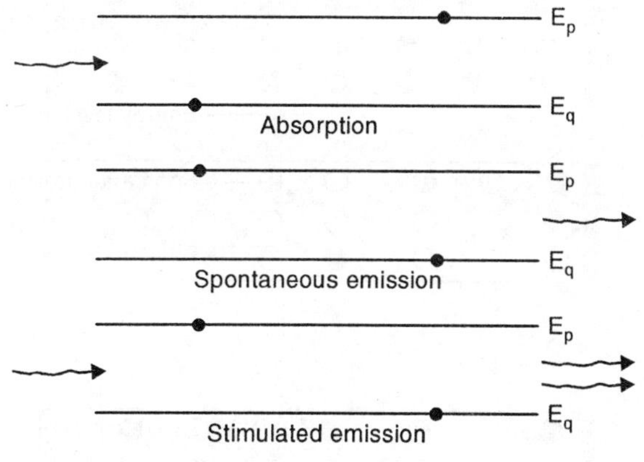

Fig. 11.2

During the period when the atom is in an excited state, if it is struck by an outside photon having precisely the energy of the one that would otherwise be emitted spontaneously, the atom can be stimulated to emit a photon thus augmenting the one given up by the excited atom. The electromagnetic wave, thus produced are identical in wavelength, phase, direction and energy with the wave that triggered its release.

Many materials can be made to undergo stimulated emission given the right circumstances. However, to build a working laser, additional conditions must be met. First, the energy source that provides the initial stimulation must be powerful enough to ensure that the majority of the atoms or molecules in the material to be lased are at their higher energy level. This condition is known as population inversion, referring to the majority population of atoms or molecules in the lasing material.

The second condition required to produce a laser is to provide a feedback mechanism. The feedback mechanism captures and redirects a portion of the coherent photons back into the active medium to stimulate the emission of still more photons of the same frequency and phase. This action is similar in operation to that of an electronic feedback oscillator.

The feedback mechanism is designed to allow a small percentage of the coherent photons to exit the system in the form of laser light. Although some of the photon are allowed to escape the system, most will still be available to maintain the amplification process through stimulated emission. The small percentage that was allowed to escape is laser light, hence the name, light amplification by stimulated emission of radiation.

11.2 PRODUCTION OF PHOTON CASCADE

A gas laser consists of a long, small-diameter glass tube that contains a gas at low pressure. Electrodes placed in the gas at both ends of the tube provides a path for electrical energy to stimulate the atoms or molecules of the gas. Before the cascade begins, the atoms in the laser crystals are in ground state shown by dots in figure below. Pumping of energy raises most of the atoms to the excited state.

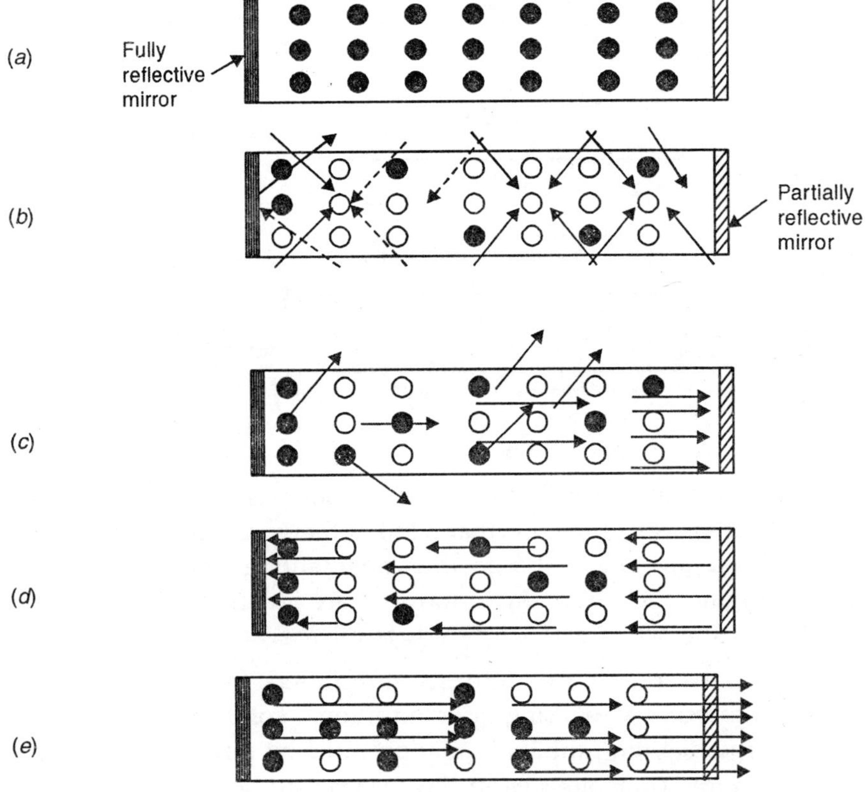

Fig. 11.3. *Production of photon cascade*

The parallel mirrors at both ends of the laser tube are placed to establish feedback mechanism for the laser resonator. One mirror is selected as 100% reflective, while other is partially reflective to provide the laser output.

When electricity is applied to the gas, it forms a plasma, and photons are emitted in all directions. The small percentage of photons that were emitted along the optical axis of the resonator are reflected by the mirrors to provide amplification. Those photons that were not emitted along the optical axis are lost from the system and removed as waste heat.

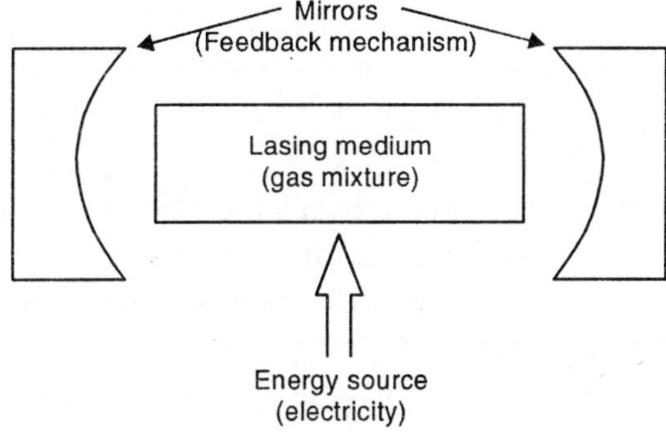

Fig. 11.4

The photons travel along the optical axis will spontaneously emit a photon from the excited atoms. These photons stimulate another atom to contribute a second photon. This process continues as the photons are reflected back and forth between the ends of the mirrors. When the amplification is large enough, some of the beam bursts out through partially mirrored end causing "lasing" action.

The output light will be very monochromatic because stimulated emission is a resonant process.

The light emitted from a laser differs from all other natural and man-made light sources and is able to perform material processing tasks because of its properties of monochromaticity, coherence, divergence and brightness.

The laser light is monochromatic means that its wavelength occupies a very narrow portion of the spectrum. This is in contrast to white light, which is composed of many different colours. The net result is that a simple lens can be used to focus and concentrate laser light to a much smaller diameter spot and a much higher intensity than with other types of light. The low divergence rate of lasers is an important factor in achieving a high intensity on the workpiece.

11.3 LASERS

Lasers used for material-processing applications must be high-power, reliable and have a relatively low acquisition cost. There are five types of lasers that meet the criteria for material processing. Four of them rely on crystals for the lasing medium (solid state) and one uses gas.

TABLE 11.1. Material Processing Lasers

Laser type	*Mode of operation*	*Active media*
Ruby	Pulsed	Solid-state
Nd : glass	Pulsed	Solid-state
Nd : YAG	CW/pulsed	Solid-state
Alexandrite	CW/pulsed	Solid-state
CO_2	CW/pulsed	Gas

Solid-State lasers

The solid state lasers use light derived from a high intensity light source as the excitation mechanism. Xenon or krypton filled flash lamps are used for pulsed lasers and krypton-filled arc lamps are used to produce a continuous (CW) beam. Lamp configurations can be either straight or helical. Straight lamps are usually preferred because they are less expensive and much easier to change when they wear out. Both types of lamps have a finite life time.

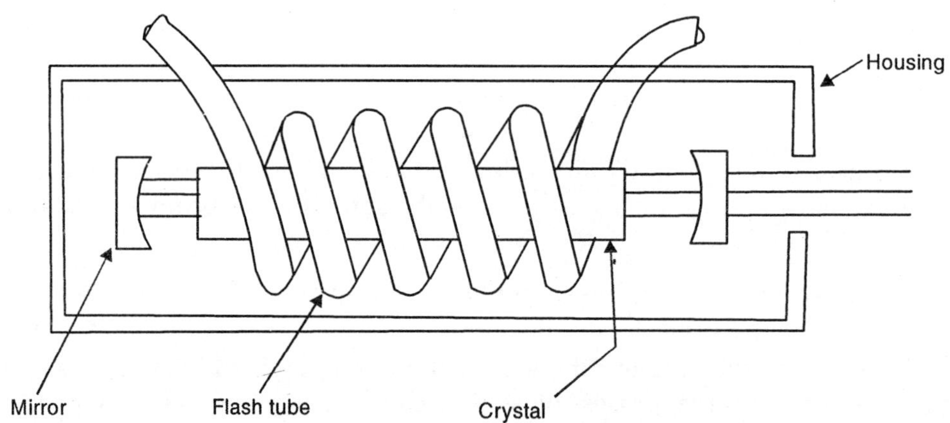

Fig. 11.5. *Solid-state laser resonator*

Two mirrors placed at both ends of the laser crystal provide the feedback of photon to sustain the lasing action.

Ruby and Nd : glass lasers cannot be operated continuously (or at pulse repetition rates faster than one or two pulses per second) because of the poor thermal properties of the rod materials and hence these lasers are strictly limited to slow, pulsed applications such as drilling and spot welding.

Gas lasers

It consists of a glass tube containing a flowing mixture of carbon dioxide, helium and nitrogen which act as a lasing medium. Direct electrical energy is used to provide the energy for stimulating the lasing medium. The gases flowing through the laser are often recirculated and replenished to reduce operating cost. The axial flow carbon dioxide laser is shown in figure 11.6.

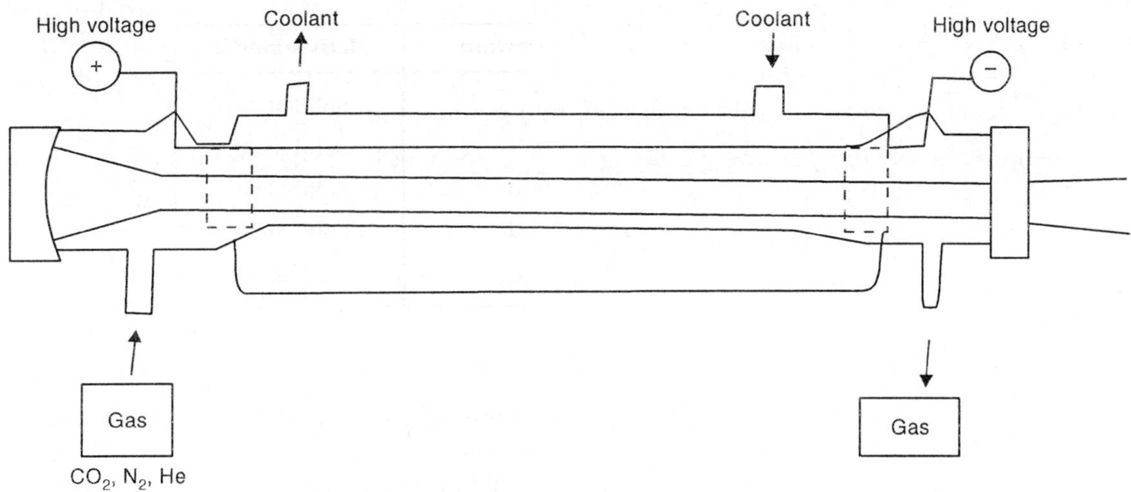

Fig. 11.6. *Axial flow carbon dioxide laser*

The power of an axial flow laser is limited to approximately 100W/m of tube length. Axial-flow carbon dioxide lasers are characterised by excellent beam quality and a high degree of stability.

When extremely compact medium-power lasers are needed or when high powers in excess of several thousand watts are required, a carbon dioxide laser incorporating a design known as transverse flow, or gas transport, is often used. The most pronounced advantage of the gas transport laser is that a large amount of power can be generated from a relatively small structure. The carbon dioxide gas mixture is continuously circulated inside the laser head. The gas mixture passes between a discharge region that is defined by electrodes positioned parallel to the optical axis. Because the gas volume in the discharge region is relatively large, the resonator mirrors can be positioned to reflect the beam through this region several times before it escapes through the output mirror.

11.4 MACHINING APPLICATIONS OF LASER

Laser has a wide range of machining applications. Material processing by laser may be categorised as follows :

(*i*) Welding

(*ii*) Material removal (drilling, trimming and evaporation)

(*iii*) Material shaping (cutting, scribling and controlled fracturing)

(*iv*) Thermokinetic change (annealing, photochemistry, grain size controls, diffusion, zone melting etc.)

Micro-welding

Lasers are used in place of more conventional joining methods when low distortion, high-speed, autogenous welds are desired. Both spot and seam welding can be performed, although only pulsed lasers capable of high repetition rates or CW lasers can be used effectively for seam welds. Parts to be welded must have a tight fit-up with a gap less than 5% the thickness of the material. Various shielding gases, such as argon or helium, are used to locally protect the weld puddle from oxidation contamination.

Depending upon the parameters being used, laser welds are accomplished by either conduction or penetration mechanics.

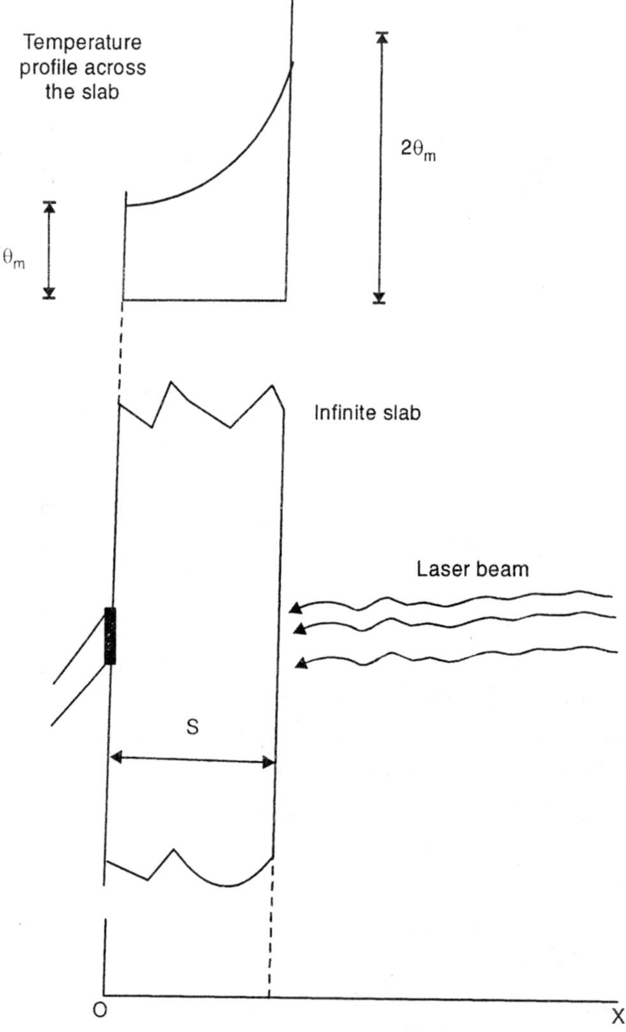

Fig. 11.7. *Welding a wire on the non-incident side of slab*

Conduction welding occurs with lower beam power densities. The depth of the weld penetration is limited by thermal conduction of the laser energy. Weld penetration is limited to a maximum of 2.5 mm. Consider an infinite slab shown in figure 11.7, where it is essential that the laser energy be delivered in such a manner as to melt but not vaporize the side of the slab on which the laser beam is incident. The temperature of this side should be above the melting point but below the vaporisation point. The energy output of the laser and the pulsing time should be so controlled as to melt the opposite side of the slab through the process of heat transfer. It is important in laser welding that both the parameters of pulse energy content and pulse length are properly designed.

It is shown by Fairbanks that if the temperature in the incident side equals twice the melting temperature, θ_m, of the material while the weld side temperature is at the melting temperature, θ_m, the pulsing time can be approximated as,

$$t \simeq 0.38\frac{s^2}{\lambda}$$

Where,

λ = thermal diffusivity, cm^2/sec

s = thickness of the slab on which the laser welding is being performed, Cm.

Thus, pulsing time should be proportional to the square of the thickness and inversely proportional to the thermal diffusivity.

Deep welds are accomplished by high-power lasers that can produce power densities high enough to actually vapourize and 'drill' a small molten channel through the workpiece. This hole, often referred to as a keyhole, is held open by vapour pressure as the beam is traversed across the workpiece. This allows the laser beam energy to be deposited deep within the material instead of only at the surface. As the hole is traversed across the workpiece, the molten wall collapse behind it and solidify, thus forming the weld nugget. This technique is referred to as penetration welding.

Laser penetration welding is capable of producing single-pass weld penetrations of 19 mm in metals.

The application of laser welding involves welding thin heat exchanger plates, used in place of resistance welding because of higher joining rate. The unique application of laser welding involves sealing electronic packages in special atmospheres.

Micro-drilling

Micro sized holes can be laser drilled in difficult-to-machine or refractory materials. This is the biggest area of application in laser machining. Laser drilling is a process for producing small holes by using either single or multiple pulses from a stationary laser beam to penetrate through the material. This technique is often referred to as laser percussion hole drilling. Percussion drilling is accomplished by placing the workpiece at or near the focal point of the laser beam. A short pulse from the laser causes a small volume of the illuminated workpiece material to both partially melt and partially vaporize. The explosive escape of the vaporising material causes most of the volume of molten material to be removed as a spray of droplets.

Round holes with diameters ranging from 0.127 to 1.27 mm can be produced with length/diameter ratios of 100 : 1.

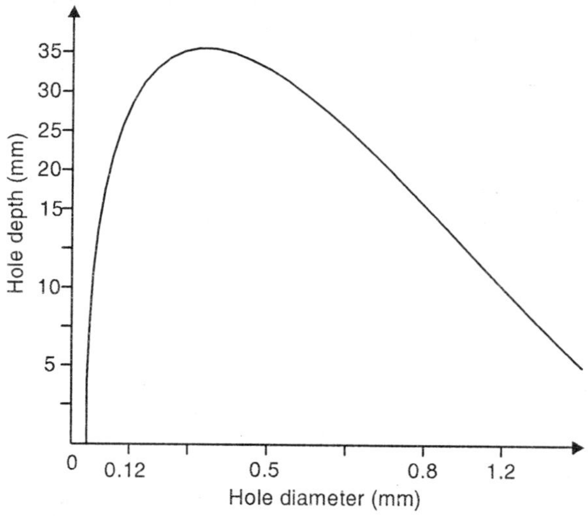

Fig. 11.8. *Drilling range for percussion technique*

The above figure illustrate the approximate range of hole sizes that are capable of being drilled by this method.

Percussion-drilled holes are characterised by a tapered, rough shape that lacks a high degree of roundness. The walls of the hole exhibit a recast layer and a heat-affected zone that can vary in thickness from 0.002 to 0.10 mm depending upon the material type, thickness and the drilling parameters.

Cutting

In common practice, when holes are larger in diameter than approximately 0.5 mm, they are in fact laser cut rather than drilled.

The process of laser cutting combines the concentrated light from the laser with a high-velocity gas jet to either vaporize or to melt and thus rapidly remove material. Cutting is accomplished at high rates of speed and without generating any cutting forces. Laser cutting is capable of piercing the workpiece at any location and can cut omnidirectionally.

In the process, the laser beam is directed to an optical assembly which houses a focusing lens and a coaxial gas jet system. The gas-jet system assists the laser beam in clearing material from the cut and also helps to keep debris from contaminating the focusing lens. The gases most commonly used for laser cutting are air, argon, or oxygen.

Solid-State transformation

Laser has been used for solid-state transformation such as selective heat treating and hardening of alloys. For laser transformation hardening to occur, the material must contain sufficient carbon (greater than 0.3%) to produce the martensitic phase that is the sole source of the hardening. To produce this transformation, the laser beam is defocused to produce a power density of only 1.5×10^2 to 1.5×10^3 W/cm^2 at the part surface. The beam is then traversed across the work surface at a rate fast enough to avoid surface melting. These conditions cause the exposed surface to be heated at rates of up to 200000°C/sec.

Heat is conducted from the surface into a thin, finite volume of metal beneath the beam. This volume of metal is rapidly heated beyond its upper critical temperature, transforming it to austenite. As the beam passes, a steady-state condition between heat input and heat conduction will be reached lasting from 0.01 to 0.6 sec. when the beam has moved on to terminate heat input, self-quenching of the heated layer occurs because of the rapid flow of heat into the cool substrate. The cooling rate can reach 20000°C/sec and results in a hard martensitic microstructure. As the rate of heat input (power density) increases, the ability of the process to produce deep hardening is reduced.

Laser hardening is most often performed by CO_2 lasers because they currently are the only type that offer power levels high enough for cost-effective hardening.

The advantages of laser hardening are derived from the ability to perform selective hardening with extremely fast thermal cycling.

11.5 PROCESS SUMMARY

Advantages

(*i*) There is no mechanical contact between the tool and the work.

(*ii*) Large mechanical forces are not exerted upon the workpiece.

(*iii*) The laser operates in any transparent environment, including air, inert gas, vacuum and even certain liquids.

(*iv*) The laser head need not be in close proximity for performing cutting and drilling operations in locations of difficult accessibility.

(*v*) The beam can be projected through a transparent window.

(*vi*) Unlike other thermal machining devices, the laser can be used with materials sensitive to heat shock such as ceramics.

Limitations

The main limitation of the laser cutting is that it cannot be used to cut metals that have high heat conductivity or high reflectivity. This means that it cannot be used to cut aluminium, copper and their alloys.

The machined area can be irregular due to off-axis modes that may be generated during laser action. The least diameter to which laser beam can be focussed depends upon on the laser beam divergence *i.e.*, quality of laser material.

REVIEW QUESTIONS

1. What is laser and how is it used to machine the materials ? Give the thermal features and analysis of the laser beam machining.
2. Make a comparison between laser beam and electron beam machining process on the basis of their applications and limitations.
3. Explain laser. List different types of lasers and their applications in industry.
4. Describe basic principle, working and general applications of Laser Beam Machining (LBM) process.
5. Describe the principle of laser beam welding process. Explain how would you select the following laser parameters for a particular application :
   (*i*) Laser type.
   (*ii*) Laser power.
   (*iii*) The laser wavelength and pulse frequency.
6. It is easy to weld with laser beam than electron beam. Justify the statement. Explain the "keyhole" technique of welding.

12

Plasma Arc Cutting

12.0 INTRODUCTION

Plasma arc cutting is a thermal material removal process that is primarily used for cutting thick sections of electrically conductive materials. High temperature plasma was first considered for cutting applications when, in the 1950's it was discovered that the arc from a tungsten inert gas (TIG) welder could be constricted to produce a much hotter plasma that could then be used for cutting. By passing the arc through a 4.5 mm diameter water cooled copper nozzle as shown below, the plasma would attain higher temperatures and disperse at a slower rate.

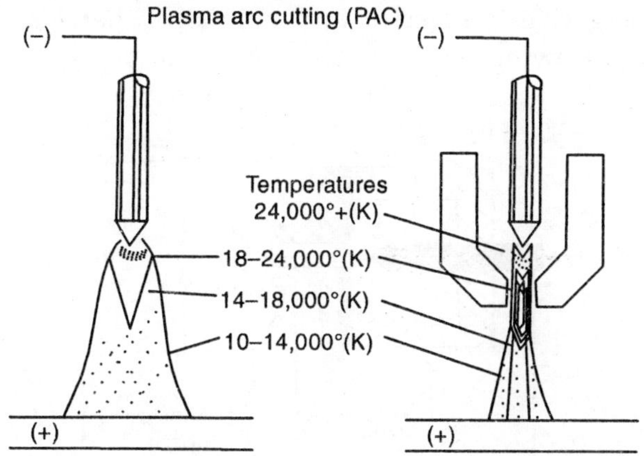

Fig. 12.1. *Temperature distribution in plasma with and without constriction*

Plasma arc cutting was introduced to industry in 1955 for cutting aluminium and stainless steels. However, it did not gain wide acceptance until approximately 1970 when new PAC techniques markedly improved the quality of the cut. In the recent years plasma arc has also been successfully used for spraying, surfacing and welding metals like aluminium, stainless steel, titanium, etc.

12.1 PROCESS PRINCIPLE

Plasma. When gases are heated to high temperature (few thousand degree), following changes take place :

(*i*) the number of collisions, elastic or inelastic, between atoms increases.

(*ii*) the gas ionizes, so that a portion of the atoms are stripped off their outer electrons, resulting in the creation of electrons and ions.

(*iii*) the electrons thus produced, in turn, collide with atoms, heat them through relaxation processes so that their thermal kinetic energy increases, excite them so that de-excitation light is emitted from the atoms and ionize them so that more electrons and ions are produced. Thus, the new matter which is called "plasma" is characterised by its ability to conduct electricity due to the presence of free charges.

The plasma is encountered in electrical discharges, such as fluorescent tubes and electric arcs, lightnings, high temperature combustion flames and the sun.

12.2 MECHANISM OF METAL REMOVAL

Earlier plasma arc cutting systems operated in what was known as the non-transferred arc mode in which power is transferred between the electrode and the nozzle, thus ionising a high-velocity gas that is streaming toward the workpiece. Because of the low thermal efficiency of this technique, all modern PAC systems now utilize the transferred arc mode. The transferred arc mode uses the electrically conductive workpiece as the positive polarity and the electrode is the negative. An electric arc, maintained between the electrode and the workpiece, heats a coaxial-flowing gas and maintains it in a plasma state.

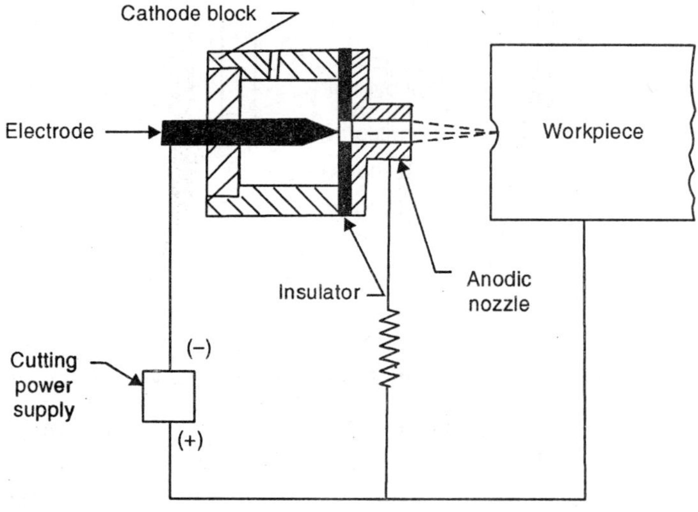

Fig. 12.2. *Schematic set-up of PAM*

The plasma arc cutting equipment can deliver upto 1000 amp at approximately 200 VDC and generate plasma temperatures upto 16000°C. The flowing gas is delivered to the torch at pressures upto 1.4 MPa resulting in a plasma velocity of several hundred meters per second. The high gasflow rate increases the efficiency of the process by adding momentum to the plasma jet to facilitate the removal of molten metal from the cut zone. The high flow rate also constricts the plasma jet and acts to provide a cool gas layer between the nozzle wall and the plasma jet. Further plasma constriction is sometimes achieved by swirling the gas as it exit the nozzle.

12.3 EQUIPMENTS

The equipment used in plasma arc cutting consists of a power supply, gas supply, cooling water system, control console and plasma torch. Of these five major items, the plasma torch is the most critical to a successful operation. The most successful torch designs that remain in common use today are the air plasma, dual gas, oxygen-injected and water-injected plasma units.

Air plasma torch

An air plasma cutting torch uses air as the gas that is ionised and performs the cutting. The constricting nozzle (made of metal) is directly exposed to the workpiece. With this design, it is possible for electricity to arc from the electrode to the nozzle and then from the nozzle to the workpiece. This undesirable arcing is known as double arcing and is the number one cause of premature nozzle failure.

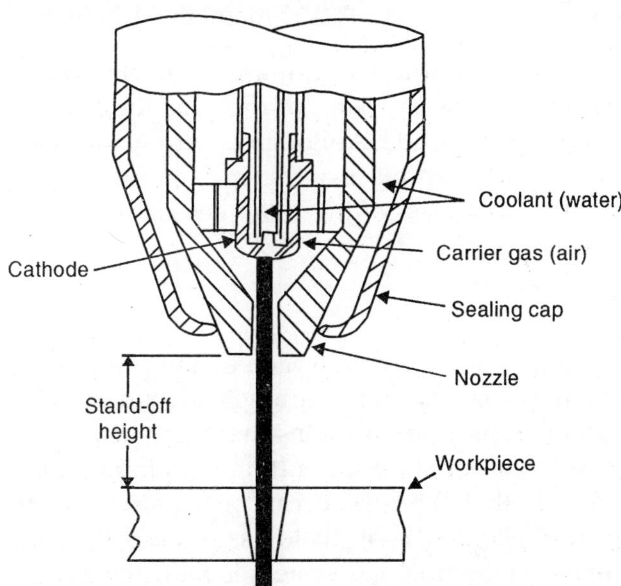

Fig. 12.3. *Air plasma torch construction*

The presence of oxygen in the air acts to support the exothermic burning of the metal and provides for very fast cuts in oxidizable materials such as steel. Other material such as aluminium, can be cut by this technique, but the cut edges are left with severe oxidation. Because of the small degree of constriction imparted to the plasma, air plasma cuts exhibit a high degree of taper alongwith rounding of the top cut edges.

Because of the elevated temperature and the presence of oxygen, a tungsten electrode would last only a matter of seconds in the air plasma torch and hence tungsten electrodes cannot be used with air plasma cutting. Instead, zirconium or hafnium electrodes are used because of their superior oxidation resistance.

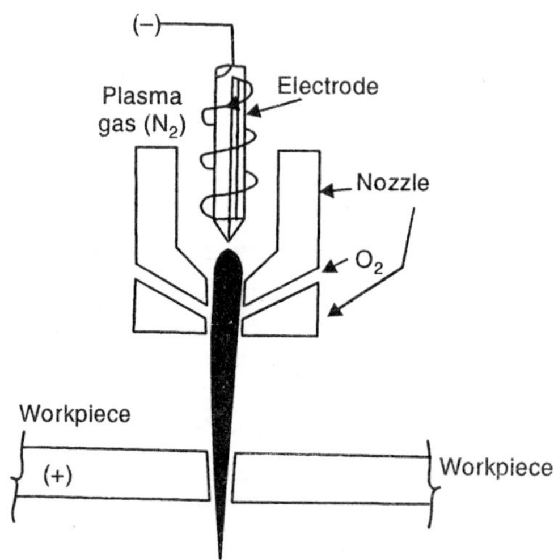

Fig. 12.4. *Oxygen-injected torch construction.*

Oxygen-injected torch

An oxygen-injected plasma torch eliminates the electrode oxidation problems encountered with air plasma torches by using nitrogen as the plasma gas and injecting the oxygen well downstream of the electrode. Although electrode life is greatly extended, nozzle life is relatively short. Oxygen-injected plasma systems are used almost exclusively for mild-steel plate cutting. While the squareness of the cut edges is rather poor, the cutting rates can be quite high with this method.

Dual gas torch

Dual gas plasma cutting system uses nitrogen as the primary gas for both generating the plasma and protecting the electrode and a variety of secondary shielding gases depending upon the material being cut. Typical applications may use such secondary gases as oxygen for steel, and an argon-hydrogen mixture for cutting aluminium. One of the advantages to using the dual gas system is that the nozzle can be recessed inside of a ceramic cup to eliminate double arcing. The relatively cool shield gas protects the ceramic cup from thermal damage. Another advantage to the dual gas system is that sharp corners can be maintained on the top side of cut edges.

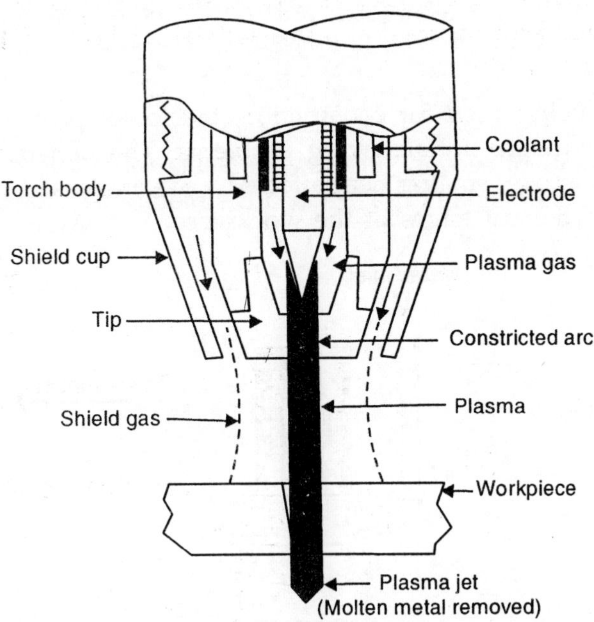

Fig. 12.5. *Dual gas plasma torch construction*

Water-injected torch

Utilizing a plasma torch with water injection results in the higher quality cuts available by PAC. With this system, water is injected either radially or as a swirling vortical cone to constrict the plasma. As the water impinges on the plasma, about 10% of the water vaporises producing a thin layer of steam which constricts the plasma. This thin boundary layer is known as the leidenfrost layer and acts to insulate the nozzle the same way as the vapour layer which causes a drop of water to glide over the surface of a very hot cooking pan without immediately vaporizing.

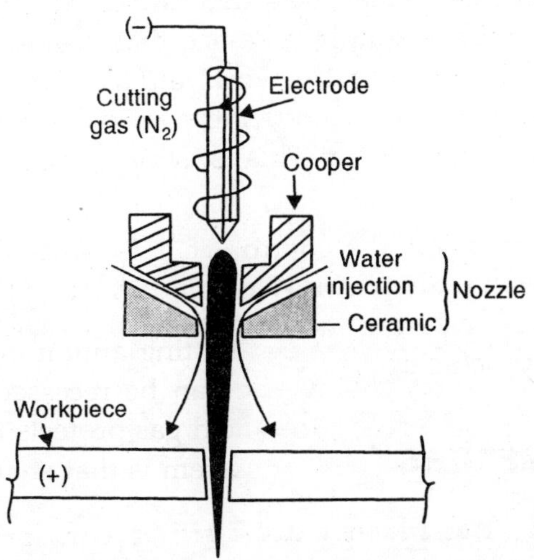

Fig. 12.6. *Water-injected plasma torch construction*

Water-injected systems use nitrogen as the plasma gas at a pressure of 1 MPa and at a flow rate of 28 L^3/min. Water is used at a pressure of 1.2 MPa and at a rate of 1.9 L^3/min to constrict the plasma.

Because water is being used for constriction, the lower portion of the nozzle can be made of ceramic to eliminate double-arcing problems. The benefits of water constriction include a reduction in smoke, smaller heat-affected zone, increased nozzle life, and limited formation of oxides on the cut edges of the workpiece.

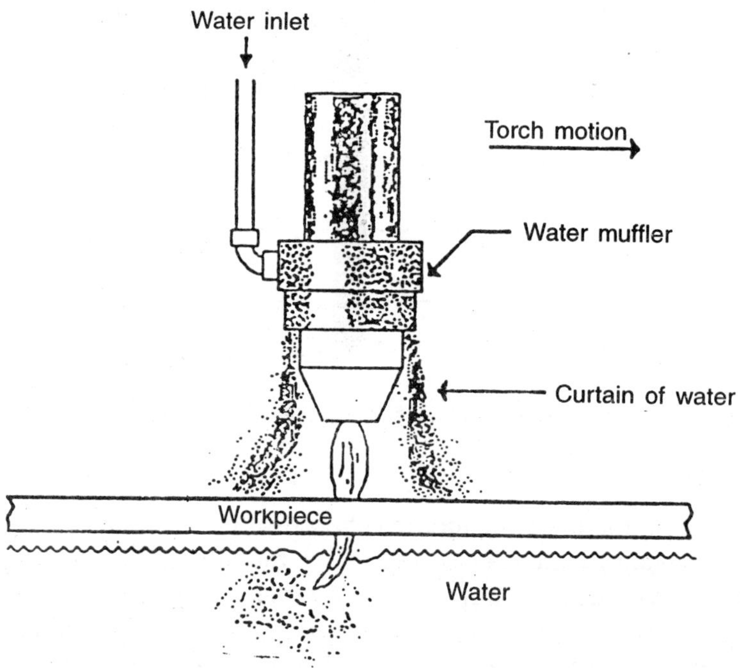

Fig. 12.7. *A water muffler is used in conjuction with a water table to reduce smoke, sparks, and noise*

Water mufflers are accessories that are often added to PAC systems for smoke and noise reduction. A water muffler is simply a device that creates a curtain of water around the plasma torch and extends down to the work surface. This has the effect of trapping smoke and dampening noise. A water muffler can also be used to absorb some of the ultraviolet light generated by the plasma. This is accomplished by mixing a dye such as nigrosine with the water.

The most effective solution for the control of noise and smoke is to place both the workpiece and the plasma torch underwater. While underwater plasma systems are available, their use is limited because of problems such as cutting thickness limitations of 38 mm and poor part visibility during processing.

12.4 PROCESS PARAMETERS

The parameters governing the plasma arc cutting can be divided into following categories :

(*i*) Those associated with the design and operation of the torch.

(*ii*) Those associated with the physical configuration of the set-up.

(*iii*) Environment in which the work is performed.

The Torch

The modified stabilized arcs came to be known as plasma torches or plasma jets because the plasma of the arc column is pushed out of the nozzle in the form of a high velocity jet. An additional feature of the plasma torches is that the anode tube is in the form of a constricting nozzle, so that further confinement and acceleration of the arc takes place, in addition to the constriction due to the stabilizing flow.

The plasma torch consists of a non-consumable cathode rod of 2 per cent thoriated tungsten and a converging anode nozzle with a suitable orifice. The two electrodes are separated by an insulator of high carbonate resin or some suitable rubber.

For vortex stabilized torches, the gas is fed tangentially through an inlet in the insulator. For sheath stabilization, the gas is fed through small ports around the cathode. Both the electrodes are water cooled.

The control provided by gas.

Modes of operation of D.C. plasma torches

Two modes of electrical connection in the torch are used.

(*i*) **Non-transferred arc mode.** The d.c. power source is connected directly across the cathode and the nozzle, so that the cathode and nozzle carry same current. In such cases, the plasma is in the form of a flame, useful for spraying, ceramic working and chemical synthesis. The hottest portion does not appear outside the nozzle. The anode dissipation is lost in useless heating of the nozzle.

(*ii*) **Transferred arc mode.** In this mode of operation, the cathode is connected directly to the negative of the d.c. source, while the anode nozzle is connected to the positive of the supply through a suitable resistor to limit the current through the nozzle to about 50 amp. The metal workpiece to be processed is then connected directly to the positive of the supply. When ignited, a pilot plasma flame is established between the cathode and nozzle which provides a conducting path for a high current constricted arc between the cathode and workpiece. Once this arc is struck the pilot flame circuit is disconnected. This method is limited to cutting, welding and hard surfacing of metals. The electro-thermal efficiency of transferred arc mode is 20–30% more than non-transferred arc mode.

Both the modes of operation use argon or nitrogen or a mixture of the two. For certain useful purposes, a percentage of hydrogen may be added.

Design of D.C. plasma torches

The plasma torch is designed to obtain maximum thermal output. The increase in efficiency not only helps in achieving better heating of the gas but also in reducing electrode losses and thereby increasing the life of the electrode. The design also ensures that the erosion rates of the electrode are kept to a minimum.

The parameters which affect the performance of the torch are cathode size, its taper near the gap, convergence of the nozzle, nozzle orifice diameter and orifice length, electrode gap and cooling of the electrode.

The cathode. The taper at the tip and diameter of the cathode affects the drop across the cathode rod. Tapered rods with slightly blunt tips are used for non-transferred applications, while for cutting applications, a flat disc slopped on the edges is used. The taper angle for N_2 is kept larger than that for argon (about 45°). The taper angle or sloping of the disc is usually smaller or equal to the convergence angle of the anode nozzle. For spraying and cutting, 10 mm dia. cathodes are used, while for welding, electrodes of 6 mm dia are used. The cathode is pressed to fit into a water cooled copper holder and brazed to it.

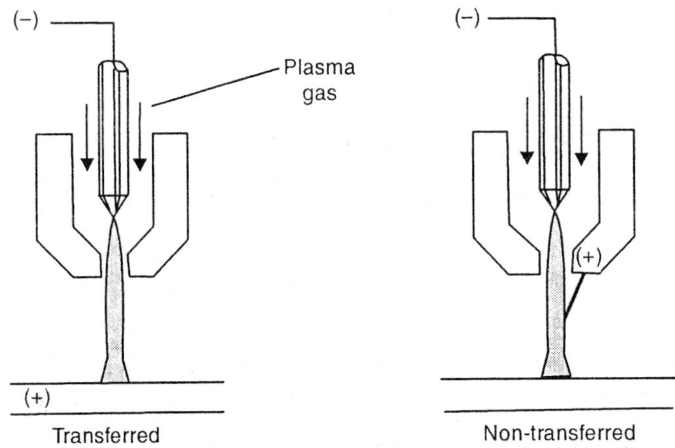

Fig. 12.8. *Transferred and non-transferred plasma arcs*

The general guidelines for designing the torch are :

(*i*) Large currents require a larger orifice diameter cathode, orifice length and electrode gap.

(*ii*) While operating at larger arc voltages, it is preferable to increase the angle of taper at the cathode tip. By the same token, while operating in the transferred arc anode (high arc voltage), flat face cathodes are used.

(*iii*) For avoiding turbulence, the edges of construction are rounded off and smaller cathode diameters are used. To stabilize laminar flow modes, a large orifice length/diameter ratio is used.

(*iv*) In any torch, the alignment between the electrodes is very essential and any misalignment would lead to undesirable arcing on the sides and non-uniform erosion on one side of the electrodes.

(*v*) Non-transferred arc modes use long throat length, while for transferred arc mode, the minimum orifice length is used.

(*vi*) The cathode taper and convergence of the nozzle should be such that the arc terminates at the tip of the cathode. This usually means that the tip is the nearest point to the anode.

(*vii*) The cooling provided to the cathode should be optimized for minimum uniform erosion without cracking the cathode and destroying the maximum stability.

(*viii*) Oxygen or compounds of oxygen in the plasma gas are detrimental to the cathode and by the same token, no water leakage into the plasma chamber from the chamber cooling.the electrodes should be allowed.

Cutting gases

The nature of application of plasma arc cutting guide the types and magnitudes of the flow rates of cutting gas. The thicker the workpiece, the greater is the gas flow requirement. The choice of the gas will depend on economics and the quality of the cut edge desired. The different gases produce cut faces of varying degrees of smoothness.

Aluminium and magnesium can be cut with nitrogen and nitrogen-hydrogen mixture or argon-hydrogen mixture. Nitrogen-hydrogen mixtures are used for cutting stainless steel up to 50 mm thick. For heavier sections, a mixture of 65 per cent argon and 35 per cent hydrogen is used.

Carbon steel can be cut most economically with gas containing oxygen but at plasma temperature pure oxygen is highly corrosive in nature and rapidly consume the electrodes.

The most suitable way to use oxidizing cutting gas is to use nitrogen as plasma gas and subsequently introduce oxygen into the plasma downstream of the electrode (dual flow). This method results in long electrode life and combines the advantages of using inexpensive gas with higher speeds obtained from an oxidizing plasma effluent.

Physical configuration

The variables such as stand-off distance, angle, depth of cut, feed and speed of the work towards the torch are playing an important role in plasma arc cutting. The feed and depth of cut determine the volume of metal removal. Figure 12.9, shows the metal removal as a function of the torch angle.

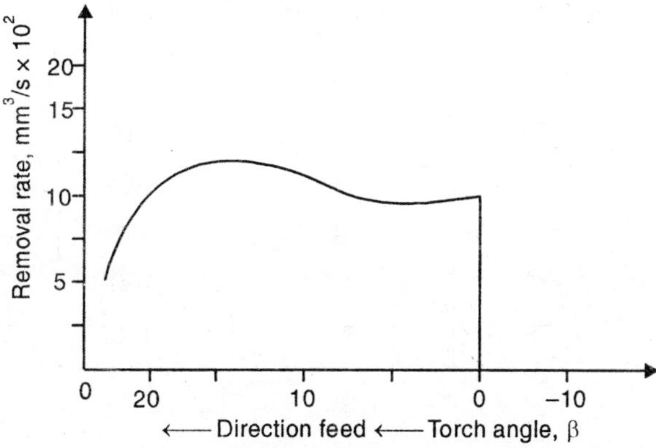

Fig. 12.9.

The operation data for aluminium is given in table 12.1. Conditions for cutting magnesium are about the same as those for aluminium. Other metals, such as

TABLE 12.1. Conditions for Plasma Arc Cutting of Aluminium

Thickness mm	Speed m/min	Orifice dia mm	Power KW	Gas flow litre/hour
6	7.5	3	60	2270 Ar – 1135 H_2 or 3700 N_2 – 900 H_2
12	5.0	3	50	1850 Ar – 925 H_2 or 3959 N_2 – 1800 H_2
25	2.25	4	80	1850 Ar – 935 H_2 or 3950 N_2 – 1800 H_2
50	0.5	4	80	1850 Ar – 935 H_2 or 3950 N_2 – 1800 H_2
75	0.375	4.5	90	3700 Ar – 2000 H_2 or 3950 N_2 – 1800 H_2
100	0.3	4.5	90	3700 Ar – 2000 H_2 or 3950 N_2 – 1800 H_2

Copper, copper-nickel alloys and cobalt base alloys can be cut, using conditions listed for stainless steel in table 12.2.

TABLE 12.2. Conditions for Plasma Arc Cutting of Stainless Steel

Thickness mm	Speed m/min	Orifice dia mm	Power KW	Gas flow litre/hour
6	5.0	3	60	4250 N_2
12	2.5	3	60	4250 N_2
25	1.25	4	80	5000 N_2
50	0.5	4	100	3700 Ar – 2000 H_2
75	0.375	4.5	100	3700 Ar – 2000 H_2
100	0.175	4.5	100	3700 Ar – 2000 H_2

There are many advantages in the quality of the cut made on stainless steel with a plasma torch as opposed to other types of cutting, particularly oxyfuel gas cutting with iron powder. In cutting with the plasma arc, there is little or no carbide precipitation, very little oxidation and therefore little contamination of the metal on the cut face.

Stainless steel, mild steel as well as aluminium in thickness upto 25 mm, can be cut at speeds upto 2.5 m/min with best quality occurring at about 1 m/min.

For mild steel cutting, using dual flow torch with N_2 as plasma gas and oxygen or air as shield gas, the cutting speed for a given thickness is double that of the conventional oxygen torch.

Work environment

The environmental variables for plasma arc cutting includes any cooling that is done on the cathode, any protective type of atmosphere used to reduce oxidation of the exposed high temperature machined surface and any means that might be utilized to spread out or deflect the arc and plasma impingement area. But in this field very little work has been done to investigate the effect of these techniques on operation.

12.5 PROCESS CAPABILITIES

Plasma arc cutting is best known for its ability to cut very thick materials, upto 150 mm thick. However, most applications lie in the thickness range of 3–75 mm. The thickness of the material being cut directly affects the cutting speed. For example, a rough estimate of the cutting speeds that can be achieved when using a 500 amp plasma arc cutting system is given by equation

$$S = 25.4/T$$

where S is in meters per minute and T is the material thickness in millimetre.

Cutting speed is also influenced by the type of material being cut. Aluminium will cut about 25% faster than steel at a given set of parameters. In general, cutting speeds for PAC are five to eight times faster than oxy/fuel cutting methods. Table 12.3 and 12.4 list PAC operating parameters for various materials and thicknesses when cut with water injected systems.

The tolerances that can be achieved with PAC are approximately ± 0.8 mm in thicknesses of less than 25 mm and ± 3 mm when thicknesses are greater than 25 mm. Taper can vary with parameters but is generally 5–7° when swirl injection is not being used. Surface finishes of approximately 5–7.5 μ for the cut edge is considered normal.

TABLE 12.3. Operating Parameters for High Current PAC Cutting

Material type	Thickness (mm)	Torch/work distance	Current (amp.)	Speed (mm/min)
Stainless-steel	100	25	1000	255
Stainless-steel	130	25	1000	150
Aluminium	100	25	900	380
Aluminium	130	25	1000	255
Aluminium	150	25	1000	180

TABLE 12.4. PAC Operating Parameters for various Materials

Material type	Thickness (mm)	Torch/work distances (mm)	Current (amp.)	Speed (mm/min)
Titanium	13	6	400	2285
Titanium	25	10	550	1270
Copper	13	6	400	1525
Copper/Ni	25	10	550	1145
Cast iron	16	6	400	2030

12.6 APPLICATION EXAMPLES

Plasma arc cutting finds widespread application in the metal fabrication and metal plate industries for shape cutting. Very high volume operations often utilize multiple-torch systems for cutting many shapes simultaneously from one plate.

The pipe industry uses PAC as a tool for preparing the ends of pipe sections before welding. In this application, the torch is mounted at a fixed angle and the pipe is rotated underneath. This results in a bevel cut on the end of the pipe. The slag build up that remains on the underside of the cut is easily removed with a chipping hammer.

Plasma arc process has also been used for other industrial applications. These major areas are :

(*i*) welding of materials like titanium, stainless steel, etc. which are otherwise difficult to weld,

(*ii*) plasma arc surfacing, and

(*iii*) plasma arc spraying.

Plasma arc surfacing

Surfacing is defined as the deposition of filler metal on metal surface to obtain desired properties or dimensions. It is usually employed to extend the life of a part which may otherwise have all the properties necessary for an engineering application, or to replace metal which has wornout or corroded away. The overlay may contribute to corrosion resistance, wear resistance, toughness or anti-friction properties.

When the overlay contributes to abrasion resistance, it is generally referred to a hard surfacing. This term is also applied to overlays used for impact resistance or low friction qualities. Here 'hard' connotes durability.

The plasma arc welding process can be used fusing thin overlays to base metals at low dilation levels. However, for weld surfacing, metal powder is used instead of wire. The powder is introduced into the effluent of the plasma arc where it is melted and bonded to the surface of the base metal. The deposited metal solidifies as a cast structure, having similar metallurgical characteristics as those of the deposits produced by the gas shielding arc welding process. The plasma arc surfacing process should not be confused with metallizing.

Plasma arc spraying

Spray coating is a process in which a surface of arbitrary thickness is obtained by spraying the previously prepared surface of the base material with droplets of a molten materials.

The process of plasma arc spraying consists of two basic operations, the preparation of the surface to be coated and spraying proper.

The surface of the parent metal is cleaned by degreasing and appropriately roughened by grit blasting.

12.7 PLASMA ARC WELDING

Plasma arc welding is an arc welding process wherein coalescence is produced by the heat obtained from a constricted arc set up between a tungsten/alloy tungsten electrode and the water-cooled (constricting) nozzle (non-transferred arc) or between a tungsten/alloy tungsten electrode and the job (transferred arc). The process employs two inert gases, one forms the arc plasma and the second shields the arc plasma. Filler metal may or may not be added. Pressure, normally, is not employed.

Process principle

Plasma arc welding is a constricted arc process. The arc is constricted with the help of a water-cooled small diameter nozzle which squeezes the arc, increases its pressure, temperature and heat intensively and thus improves arc stability, arc shape and heat transfer characteristics.

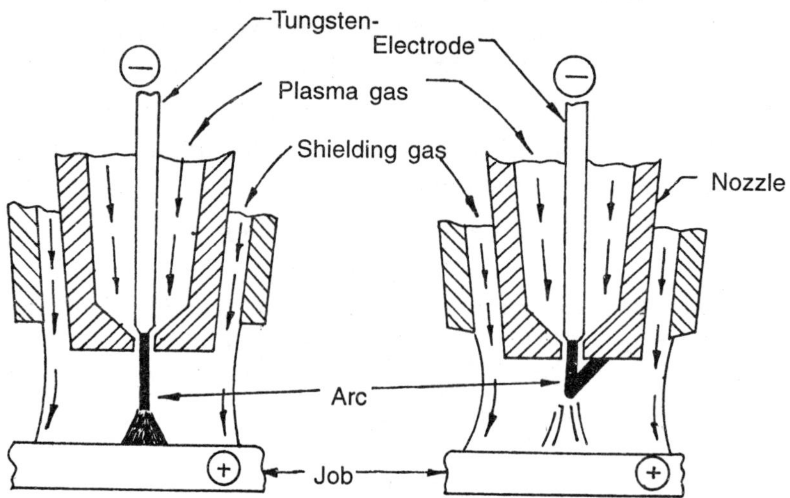

Fig 12.10. *Concept of transferred and non-transferred arc processes*

There are two types of plasma arc welding processes :

(*i*) Non-Transferred Arc Process. The arc is formed between the electrode (–) and the water cooled constricting nozzle (+). Arc plasma comes out of the nozzle as a flame. The arc is independent of the workpiece and the workpiece does not form a part of the electrical circuit. Just as an arc flame, it can be moved from one place to another. The non-transferred arc plasma possesses comparatively less energy density as compared to a transferred arc plasma and it is employed for welding and in application involving ceramics or metal plating.

(*ii*) Transferred Arc Process. The arc is formed between the electrode (–) and the workpiece (+). A transferred arc possesses high energy density and plasma jet velocity. For this reason, it is employed to cut and melts metals. For initiating a transferred arc, a current limiting resistor is put in the circuit which permits a flow of about 50 amps between the nozzle and the electrode and a pilot arc is established between the electrode and the nozzle. As the pilot arc touches the job, main current starts flowing between electrode and job; thus igniting the transferred arc. The pilot arc initiating unit gets disconnected and pilot arc extinguishes as soon as the arc between the electrode and the job is started. The temperature of a constricted plasma arc may be of the order of 8000–25000°C.

Equipments

The plasma arc welding system consists of power supply, high frequency generator, plasma torch, shielding gas, voltage control, current and gas decay control and fixture.

(*i*) Power supply. A direct current power source (generator or rectifier) having dropping characteristics and open circuit voltage of 70 volts or above is suitable for plasma arc welding. Working with helium as an inert gas needs open circuit voltage above 70 volts. This higher voltage can be obtained by series operation of two power sources, or the arc can be initiated with argon at normal open circuit voltage and then helium can be switched on.

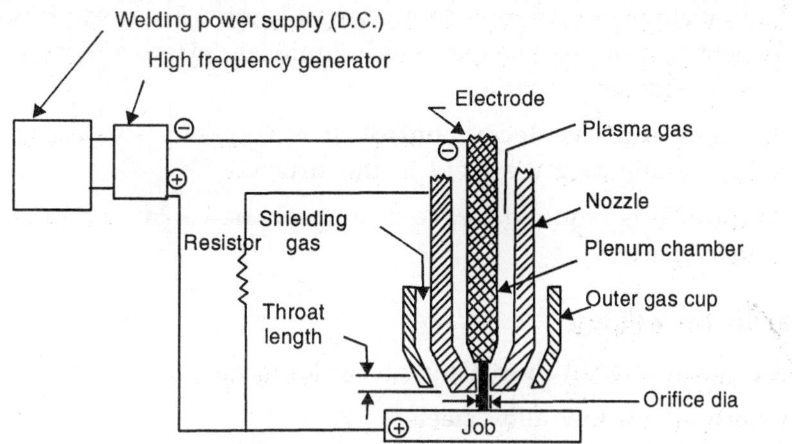

Fig. 12.11. *Plasma arc welding system*

The welding parameters for plasma arc welding are current 50–350 amps, voltage 27–31 volts, gas flow rates 2–40 litres/minute (low range for orifice gas and higher range for outer shielding gas), DCSP is normally employed except for the welding of aluminium in which case water-cooled copper anode and DCRP are preferred.

(ii) High frequency generator and current limiting resistors are used for arc ignition. Arc starting system may be separate or built in the system.

(iii) **Plasma torch.** It is either transferred arc or non-transferred arc type. It is hand-operated or mechanised. At present, almost all applications require automated system. The torch is water cooled to increase the life of the nozzle and the electrode. The size and the type of nozzle tip are selected depending upon the metal to be welded, weld shape and desired penetration height.

(iv) **Shielding gas.** Two inert gases or gas mixtures are employed. The orifice gas at lower pressure and flow rates forms the arc plasma. The pressure of the orifice gas is intentionally kept low to avoid weld metal turbulence, but this low pressure is not able to provide proper shielding of the weld pool. To have suitable shielding protection, same or another inert gas is sent through the outer shielding ring of the torch at comparatively higher flow rates.

Most of the materials can be welded with argon, helium, argon + hydrogen and argon + helium, as inert gases or gas mixtures. Argon is very commonly used. Helium is preferred where a broad heat input pattern and flatter cover pass is desired.

For cutting purposes a mixture of argon and hydrogen (10–30%) or that of nitrogen and hydrogen may be used. Hydrogen, because of its dissociation into atomic form and thereafter recombination generates temperatures above those attained by using argon or helium alone.

(*v*) **Voltage control.** Voltage control is required in contour welding. In normal key hole welding a variation in arc length upto 1.5 mm does not effect weld bead penetration or bead shape to any significant extent and thus a voltage control is not considered essential.

(*vi*) **Current and gas decay control.** It is necessary to close the keyhole properly while terminating the weld in the structure.

(*vii*) **Fixture.** It is required to avoid atmospheric contamination of the molten metal under bead.

Base Metal to be welded

The base metal welded by plasma arc welding are :

(*i*) Carbon and low alloy steels

(*ii*) Stainless steels

(*iii*) Copper alloys

(*iv*) Nickel and cobalt base alloys

(*v*) Titanium alloys

(*vi*) Aluminium alloys

Advantages

The plasma arc welding has following merits :

(*i*) Stability of arc.

(*ii*) Uniform penetration.

(*iii*) Simplified fixtures.

(*iv*) Rewelding of the root of the joint saved.

(*v*) It is possible to produce fully penetrated keyhole welds on pieces upto and about 6 mm thick with square butt joint.

(*vi*) Excellent weld quality.

Disadvantages

(*i*) Infra-red and ultraviolet radiations necessitates special protection devices.

(*ii*) Welder need ear plugs because of unpleasants disturbing and damaging noise.

(*iii*) More chances of electrical hazards are associated with this process.

(*iv*) The process is limited to metal thickness of 25 mm and lower for butt welds.

(*v*) Inert gas consumption is high.

Applications

(*i*) Single run autogenous and multirun circumferential pipe welding.

(*ii*) In tube mill application.

(*iii*) Welding cryogenic, aerospace and high temperature corrosion resistant alloys.

(*iv*) Nuclear submarine pipe system (non-nuclear section, subassemblies).

(*v*) Welding steel rocket motor cases.

(*vi*) Welding of stainless steel tubes (thickness 2.6 to 6.3 mm).

(*vii*) Welding titanium plates upto 8 mm thickness.

(*viii*) Plasma torch can be applied to spraying, welding and cutting of difficult to cut metals and alloys.

12.8 PROCESS SUMMARY

Advantages

(*i*) Cuts any metal.

(*ii*) Faster than ony/fuel by 5–10 times.

(*iii*) 150 mm thickness ability.

(*iv*) Applicable to automation.

(*v*) Omnidirectional-cutting ability.

Disadvantages

(*i*) Large heat affected zone.

(*ii*) Taper.

(*iii*) Sharp corners difficult to produce.

(*iv*) Smoke and noise.

(*v*) Burr often results.

REVIEW QUESTIONS

1. Discuss some of the important considerations in the design of a plasma torch. What are the essential differences between a cutting and a welding torch ?
2. Discuss the factors that influence the quality of the cut in Plasma Arc Cutting.

3. Explain what is meant by nontransferred and transferred mode of plasma arc. What are the advantages of each ?
4. What are the essential differences between plasma spray and only-gas spraying ? Discuss in detail.
5. How is plasma produced ? Why is plasma welding more effective for high melting alloys ? Explain.
6. Explain how would you select the plasma-gas, speed of welding and the stand-off distance to be used in a plasma welding process.
7. Explain why a nozzle in Plasma Arc Machining (PAM) process is designed carefully. Give limitations of PAM.

❏❏❏

Metallizing Process

13.0 INTRODUCTION

Dr. Ulrich in 1910 designed a nozzle that permitted an annular flame to heat the (powder) particles as they were propelled onto a surface. His further work led to the utilization of a wire in place of the powder and the equipment became known as wire metallizing gun.

Metallizing or metal spraying terms denoted the application of metallic coatings to a base metal. Later on, when it became possible to deposit ceramic materials also, by this process, a more general term known as flame spraying was coined. Flame spraying applied to the deposition of both metallic and non-metallic materials with the help of a combustion flame.

With the developments of plasma-generating devices and electric arc units, flame spraying could no longer comprise all types of deposition and another term known as thermal spraying entered the picture. Thermal spraying covers all deposition processes *e.g.*, metallizing, flame spraying, electric arc spraying, plasma spraying and detonation spraying by which surfacing materials are made molten or semi-molten by a heat source and propelled onto the base metal surface by a rapidly moving gas stream.

Metallizing as a process normally includes the preparation of base material, the spraying on the metal and finally finishing the surface by grinding.

13.1 PROCESS PRINCIPLES

The process consists of two basic operations :

(*i*) Surface preparation

(*ii*) Spraying operation

Surface preparation

The surface preparation is of paramount importance for the perfect adhesion of the coating to the parent material. For this the base metal surface is cleaned of all foreign substances such as oxides, oil, water etc. Oxide and other coatings may be removed by using mechanical methods such as machining, grinding, sanding etc. while grease and oil can be chemically removed from the base metal surface.

Afterward the base metal is roughen by forming keys or openings on it so that the sprayed metal penetrates in them thereby resulting in a bonding or keying action. The methods employed for this purpose are :

(*i*) Rough threading

(*ii*) Grit blasting

(*iii*) Electric bonding

(*iv*) Studding

(*v*) Sanding or grinding

(*vi*) Rough machining or grooving

Shafts may be rough threaded to ensure a good bond between the shaft and sprayed metal. Grit blasting is perhaps the most common method of surface preparation especially for large surfaces. Medium grade clean sharp sand, steel grit and aluminium oxide are used as abrasive materials. When surfaces is too hard to roughen, the electric bonding may be used to deposit a bonding coat of nickel alloy on the surface. In this process, a nickel or nickel alloy electrode is used to fuse small irregular particles to the surface which is to be metallized. Electric bonding is not suitable for thin deposits, it may produce slight distortion.

Studding implies placing studs (a machine screw or a rivet) on the base metal surface to increase bonding qualities of the sprayed metal.

Rotary hand grinders are used to prepare hard alloy surfaces that cannot be effectively under cut or rough threaded.

Grooving involves cutting a series of parallel grooves, using a rounded grooving tool. The lands between the grooves may then be rolled over with a knurling tool.

Since almost any oxide-free metal will adhere to molybdenum, a layer of it uniformly sprayed onto the base metal to produce a surface rough enough to receive a metallizing spray. This method of surface preparation is known as spray bonding.

Flame spraying (metallizing)

Basically, there are two types of equipments used for metal spraying. One is the metallic gun which consists of a gas torch with a hole in the centre of the tip for the wire, a small air turbine and gears to feed in the wire through the tip into the flame as fast as it melts and an air cap around the torch tip and nozzle which supplies a blast of air to atomise the molten metal and deposit in on the prepared surface. In other method, powdered metal is fed from a container through a rubber hose to the spray gun and out through the centre of the flame, similar to the wire gun. In this case metal is already in the atomised form and hence air needed is just sufficient to deposit the molten metal on the surface being coated. The metallic gun using metal in the wire form is most commonly used. The wire is fed into the spray gun at a definite rate melted by an oxyacetylene flame and then blown on the surface being coated, by compressed air.

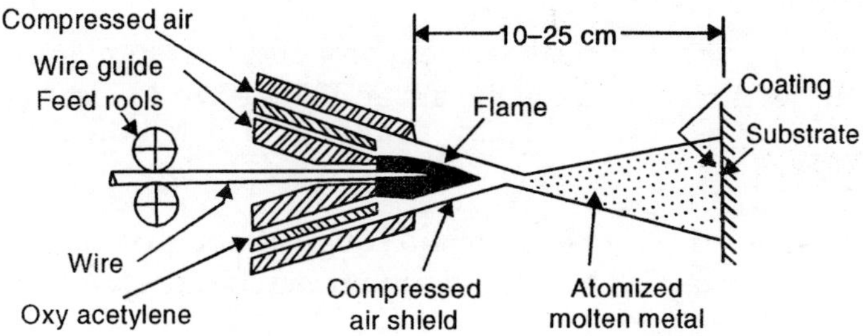

Fig. 13.1. *Flame spraying (wire metallizing)*

A wire type metal spraying system is probably the most commonly used metallizing process employed today. The wire metallizing is a fast and low-cost method of applying a wide variety of metal coatings, such as carbon steel, stainless steel, brass and bronze over a grit blasted or nickel-aluminide-under coated substrate. Such coatings are often used for shaft build up and general restoration of worn or mismatched parts.

The advantages of the powder-fed process is its ability to apply a larger variety of metal, ceramic and cermet (ceramic/metallic mixture) materials. As compared to wire-metallizing equipment, powder spraying equipments is simpler and lighter in weight. One problem experienced in powder spray process is the difficulty of maintaining a steady feed of powder. Moreover, deposition efficiency is low.

Electric arc spraying

In some cases the adhesion of a flame sprayed coating is unsatisfactory because it is impossible to attain a sufficiently high temperature in the particles on impact. This problem has been overcome by electric arc spraying, in which maximum temperature of about 4000°C have been achieved in the arc.

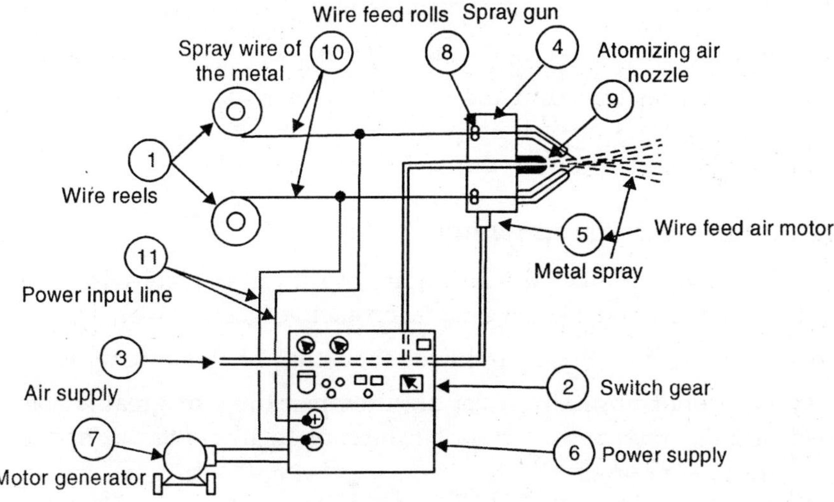

Fig. 13.2. *Electric arc spraying*

An electric arc is maintained between two wires of the metal which is required to be deposited on the workpiece surface.

The metal melted by the arc heat is then propelled or sprayed onto the workpiece surface by using an air jet directed across the arc and at the workpiece surface to be sprayed.

Arc spraying operation

The surface to be sprayed must be clean and rough although considerably less roughness is required for electric arc spraying. Sand and aluminium oxide should not be employed for preparing the base metal surface.

The two metal wires are advanced by the feed mechanism under full electric current (wires pass through the electrical wires guides and become energised) and they touch at a low contact pressure and a small point-like contact surface.

Because of the high density of the current and the extreme heat of the contact surfaces, an arc strikes immediately. The arc creates a molten condition at the ends of the electrode wires and the molten particles are swept off at high velocities by a gas (*e.g.*, nitrogen) or compressed air jet. The vibrations of the wire ends as they make and break contact assist the spraying action. Thus, the molten particle stream is impinged upon the workpiece surfaces being sprayed, resulting in the deposition of a high quality coating.

Advantages of electric arc spraying

(*i*) Arc spraying is approximately 70% less costly than flame spraying. One of the reasons for this is its high rate of deposition in spraying steel, copper and copper alloys, zinc, lead, aluminium and white metal from wire.

(*ii*) No combustion gases are required, thus eliminating all problems and dangers associated with these gases.

(*iii*) Arc spraying provides a high level of thermal efficiency. In this process, the two wire ends are resistance heated between the short circuit and contact points with 40 to 50% of the energy produced in the electric arc.

(*iv*) Arc spraying requires less surface preparation than flame spraying.

(*v*) Arc sprayed particles and the coating layer is oxidised as much as 50 to 70% less than flame sprayed particles, resulting in higher resistance to tensile, compressive and bending stresses.

Disadvantages of electric arc spraying

(*i*) Considerable quantities of carbon are burnt out of steel wires and a loss of Mn and Si can be expected because a protective gas is not employed.

(*ii*) The arc spraying velocity is lower than that of flame spraying.

(*iii*) Coating do not normally resist high temperatures or great surface pressures or load applied over small areas because of their limited adhesion. Neither can they resist stressing.

(*iv*) Process creates fumes, dust and odours around the work.

Plasma arc spraying

The surface to be sprayed must be cleaned of dust, oil and other foreign matter. After that, the surface should be roughened by (steel) grit blasting or acid etching.

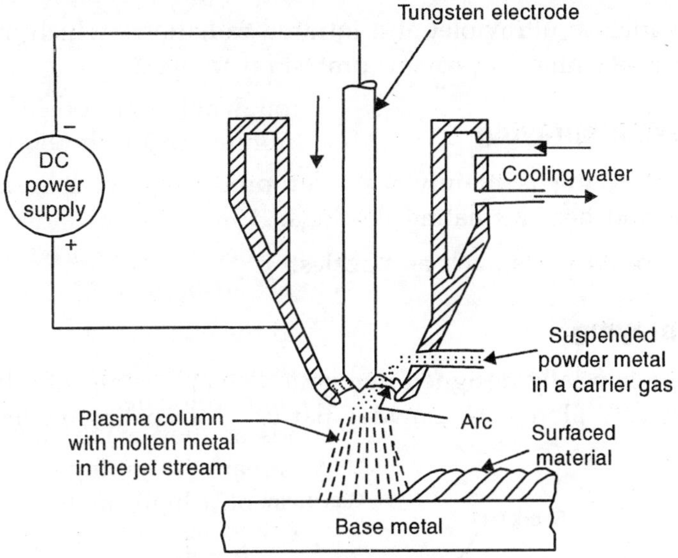

Fig. 13.3. *Plasma-arc spraying*

In plasma arc spraying, a non-transferred, direct current, high intensity arc is struck between the gun body, which acts as an (Cu) anode and a thoriated tungsten cathode (electrode).

An inert gas, *e.g.*, argon is passed through this arc, where it is heated to the plasma state (20000° to 30000°F) and accelerated to supersonic speeds. Argon, heated by the arc, escapes out of the nozzle (*i.e.*, copper anode) as a hot, brightly shining flame having a length of about 3–7 cms.

Powder metal is introduced into the plasma stream which reaches the molten stage much more quickly than in flame spraying. The powdered metal is then carried to the workpiece surface in a molten state by the high velocity of the plasma jet.

Arc current	—	80 – 800 amps
Voltage	—	50 – 500 volts
Powder metal size	—	upto 5 micron
Distance of gun from work	—	50 to 150 mm.

Advantages of plasma arc spraying

(*i*) Any base material or substrate can be coated with the plasma spray process.

(*ii*) High bond strength is obtained.

(*iii*) The coating is stronger and denser.

(*iv*) Heat distortion problems are minimum.

Disadvantages of plasma arc spraying

(*i*) There is a high noise level ranging around 100 dB, this may damage the inner ear.

(*ii*) There is intense ultraviolet and infrared radiations, which may cause sunburn, making it essential to provide protection to eyes.

Application of plasma spraying

(*i*) Plasma spraying is mainly used for producing wear resistant, temperature resistant and heat insulating coatings.

(*ii*) Making rocket parts such as nozzles.

Detonation gun spraying

It makes use of a specially designed detonation chamber into which metered amounts of oxygen and acetylene, alongwith powder particles of coating materials suspended in nitrogen are kept.

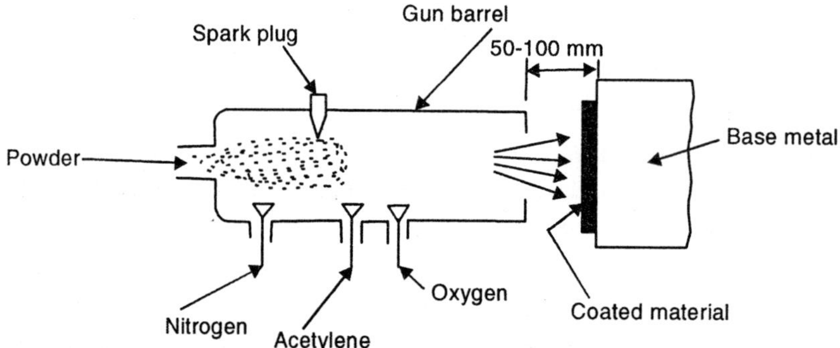

Fig. 13.4. *Detonation gun spraying*

The gaseous powder mixture is then ignited with the help of a spark plug. The resulting detonation produces a high velocity shock front which travels down the length of the barrel at ten times the speed of sand, accelerating the powder particles which have been heated to a plastic state by the detonation. The wave front caused by the explosion forces the particles out of the barrel and on to the base metal.

The high kinetic energy of each particle is converted to additional heat upon impact with the base metal surface, thereby producing a band that is both metallurgical and mechanical in nature.

Unlike metallizing and plasma torches which operate, continuously, the detonation gun fires 4 to 8 times a second, thereby forming a laminar coating on the workpiece. Successive detonation can build up coating to a maximum thickness of 0.75 mm.

Auxiliary cooling is employed to keep workpiece temp. between 65 and 150°C. Because of the extreme noise caused by the firing of gun, the process has been completely automated and the gun is usually located within concrete sound–absorbing walls and is aimed and fired by remote control.

Applications

A variety of coatings including tungsten carbide, aluminium oxide and chromium carbide may be applied with this process. The substrate is only metallic, because non-metallic surfaces may be eroded by the high velocity of the particles.

13.2 PROCESS SUMMARY

Advantages of sprayed coatings

By spraying metal on the parent metal, the original properties are changed. There is an increase in porosity, compressive strength and hardness, but the tensile strength decreases. Also the wearing quality of sprayed metal is good. There is no distortion and no internal stresses are developed in the parts.

Disadvantages of sprayed coatings

All sprayed metal is brittle and therefore is limited to use on machine elements where it is not necessary for the coating to add strength.

13.3 APPLICATIONS OF METAL SPRAYING

(*i*) **Corrosion-Protection.** The most extensive use of metal spraying is the application of aluminium and zinc into iron and steel. Colorising is frequently accomplished by spraying of 0.125 mm aluminium on steel and then heating the part to form iron aluminium compound on the surface.

(*ii*) **Hard and Noble Surface.** One big application of metallizing is the application of special metal surface on large masses of less costly metal *e.g.*, a big shaft required to be corrosion resistant need not be made completely of stainless steel or build up worn out surface of machine components. It might be made first of any metal then metallized with noble metal. Similarly the metallizing may be applied for abrasion or wear resistance, corrosion protection and electrical or magnetic properties etc.

(*iii*) **Soldering Surface.** Sprayed copper is frequently used on non-metallic parts when it is desired to attach parts by soldering.

(*iv*) **Electrical Conductivity.** Conductive coating of copper and silver can be obtained on poor conductor *e.g.*, by spraying copper on most of carbon brushes for motors and generators for better electrical connection.

(*v*) **Thermal Conductivity.** Sometimes it is desirable to have surface to carry away heat from hot spots on poor conductors. In this case copper can be sprayed.

(*vi*) Its other applications are in having decorative films, reflecting surface and special metal forms.

(*vii*) Mechanically size a component that has not been produced in correct specification.

REVIEW QUESTIONS

1. In the present scenario, what is the importance of metallizing process ?
2. What are the different steps involved in the metallizing operation ? Discuss the various methods employed for surface preparation.
3. Discuss the plasma Arc spraying process with a neat sketch.
4. What is difference between plasma arc praying and detonation gun spraying ?
5. Mention the various applications of metallizing process.

❑❑❑

High Energy Rate Forming

14.1 INTRODUCTION

Since early man's first use of a hand-held hammer, progressive increases have been made in the energy that could be stored and imparted to a workpiece. First he developed the gravity drop hammer, then the steam driven hammer and finally, the electrically powered mechanical and hydraulic presses. These increases in energy were derived mainly by the development of means for manipulating larger and larger masses. However, forming velocities only gradually increased. Thus, most conventional forming processes deform metal by using massive tools moving at relatively low velocities.

High-energy rate forming (HERF) refers to processes that form parts at very high velocities and extremely high pressures. A more accurate name for the process might be "high velocity forming" because the most distinguishing feature is the high forming velocity rather than the expenditure of large quantities of energy. All of the HERF processes involve a short, forming energy input usually of microsecond duration.

The energy levels employed in HERF vary widely depending upon the requirements of the application. While as little as 10 joules of energy may be required to form a small transistor enclosure, a large workpiece such as a tank turret may require a forming energy in excess of 1×10^9 joule.

HERF processes make use of a short burst of energy that is transmitted to the part surface through a medium such as water or air. The resulting force or shockwave acts to force the part into a die cavity that has the desired shape of the finished part. Forcing the part into a die in this manner drastically reduces die costs because the need for a male die is completely eliminated.

Materials being formed at a very high velocity behave almost like a fluid; they can be formed beyond their usual limits and maintain excellent dimensional control. Materials that are difficult to form by conventional methods because of springback will undergo large amounts of plastic deformation with almost no springback when HERF is used. In addition, HERF will typically improve material properties by improving the ultimate tensile strength and producing a more uniform strain throughout the part.

High velocity forming processes began to make their mark and grow in application by 1960. Space age requirements called for an increase in the size of the workpiece, the use of tougher and more highly heat-resistant materials, and placed a premium on flexibility in making design changes. All these factors have changed explosive forming from an old laboratory curiosity to an industrial production process competing in certain specific areas both economically and in terms of product quality, with the conventional processes.

DEVELOPMENT OF SPECIFIC PROCESSES

Explosive forming

The earliest applications of explosives to metal-working was made by C.E. Munroe, who discovered how to concentrate the energy of high explosives. In 1888, he described the engraving of iron plates by imprinting a design on a block of explosives or by interposing a stensil between the explosive and the metal plate.

Later on in 1941, E.I. du Pont de Nemours and company had produced explosive rivets for commercial sale. These rivets contained a heat-sensitive propellant type charge which could be fired by application of heat after placement of the rivet in the assembly. The shank end of the rivet would expand from the pressure of the burning powder and effect the desired riveting action.

Probably the first use of explosives for forming which could be considered a production operation occurred in 1950, when the Moore company reported the forming of some large electric exhaust fan hubs.

The advent of the Space Age finally provided the impetus in about 1955 for the development of the explosive forming process on a broader commercial scale. Manufacturing methods were needed for low-volume production of large, often complex parts of tough metals. Close tolerances were required. The use of explosives provided a means for fabricating large structures without first having to build either large, precision machine tools or massive, complicated matching die sets. It reduced the lead time for tooling and equipments.

Electrohydraulic forming

The ability to generate high-intensity shockwaves by discharging stored electrical energy across electrodes submerged in a liquid medium has been recognized for a great many years. In 1936, Suits explained that "high-intensity sparks are the most efficient means known of converting electrical energy into sound. Spark discharge in a liquid was pointed out as a potential source of power in 1944. In 1953, first time Early and Dow of USA used this method to punch holes in 1/16" steel plates

Magnetic forming

One of the earliest developments in producing a short-duration, high intensity magnetic field was reported by Kaptsa in 1924. A special storage battery was used as the energy source. Wall is often credited as being the first to use the charged capacitor for this purpose. Harvey and Brower, around 1958, demonstrated the application of a magnetic system for forming metals. The first commercial magnetic forming machine was marketed in 1962.

SELECTION OF TECHNIQUES

Explosive, propellant, combustible gas, electrohydraulic and magnetic forming methods are all applicable to sheet metal forming and to a first approximation, may be compared functionally on an equivalent energy basis. Economic factors, however, generally determine which method is most favourable for a particular application and these factors must be carefully analysed.

High explosives are probably the most versatile of the energy sources for high velocity forming processes. High explosives lend themselves to forming by open as well as closed-die techniques. More important, there are almost no limits to size of parts that can be formed. Another advantage to explosive forming is the variety of forms in which high explosives may be used as the energy source for forming, for example, special charges, pellets, cords, sheet and liquid. Thus, the optimum configuration of the energy source for a given forming operation can be readily provided.

Propellant powdered metal forming methods, although they have successfully been applied to production runs, require the use of relatively more complex and more expensive closed-die systems. The use of combustible gases suffers the same disadvantages in die costs and possesses the same potential advantage in heat utilization. It is not easy to evaluate either of these processes as economically superior to open die explosive forming techniques except for those circumstances in which thermal requirements or complex part geometry dictate comparable die complexity for the explosive technique as well.

Electrohydraulic systems are inherently more capable being converted to automatic production processes than explosive systems. However, the high production rates or high repetition rates of the discharge of electrohydraulic systems can impose severe power supply requirements. Although the operating costs of electrical storage systems are low, the capital cost is relatively high. Thus, electrohydraulic forming appears to offer the greatest advantage in the small to intermediate range of part sizes that require moderate energy levels.

Magnetic techniques are also capable of automation with reservations similar to those for the electro-hydraulic systems. A magnetic field of 1/2 megagauss has approximately the same energy density as a high explosive. Creating such intense fields over appreciable volumes, even for times of the order of 0.1 sec. is very difficult and expensive. The power supply problem may be much less severe because common applications of the magnetic method do not involve stored energy levels as high as those considered for the electrohydraulic method. A key problem in magnetic forming is the proper design of the coil, because it is normally subjected to the same forces as those exerted on the workpiece.

14.2 COMPARISON OF CONVENTIONAL AND HIGH VELOCITY FORMING METHOD

Advanced designs of aeronautic and aerospace vehicles have presented the sheet metal forming industry with the need to form large or complex structures from the high-strength, thermal-resistant alloys. These needs have resulted in the development and growth of new

TABLE 14.1. Summary of High Velocity Forming Methods

		Explosive Forming			Pneumatic Mechanical	Electric Discharge		
		High Explosives	Propellants	Gas Mixture	High-pressure gas	Exploding wire	Spark discharge	Magnetic field
Source characteristics	Method of energy release	Chemical detonation	Chemical burring	Chemical burring	Quick-release valve	Vaporization of wire	Ionisation of medium	Collapsing magnetic field
	Energy transfer medium	Water, air, sand	Air, water	Gas	Air	Water, air	Water, air	Air
	Pressure wave velocity (ft/sec)	4,000–25,000	1,000–8,000	1,000–8,000	50–200	20,000	20,000	10,000–20,000
	Pressure wave duration	Micro-seconds	Milli-seconds	Milli-seconds	Milli-seconds	Micro-seconds	Micro-seconds	Micro-seconds
Adaptability	Location of facility	Remote	Separate	Separate	Unrestricted	Separate	Separate	Separate
	Parts diameter limits	10ft	5ft	5ft	Present – 1.5ft Future – 3ft	Present – 5ft Future – 10ft	Present – 3ft Future – 5ft	Present – 1ft Future – 4ft
	Shape versatility	Good	Good	Fair	Excellent	Good	Good	Fair
	Uses or technique reported to date	Blanking, coining, powder compacting, embossing, sizing, drawing, expanding	Bulging, compacting, driving, machining	Bulging, stretching, sizing	Compacting, drawing, extruding, forging, upsetting	Bulging, stretching, sizing	Expanding, bulging, sizing stretching, coining, embossing	Swaging, joining shrinking
Process Selection Factor	Main advantage	Large part size	Hand tool	Pressure control	Precision forgings	Reproducibility	Operating ease	
	Capital cost	Low to medium	Low	Medium	Medium to high	Medium to high	Medium to high	Medium to high
	Tooling cost	Low	Medium	High	Medium	Medium	Medium	High
	Operating cost	High	Low to medium	High	Low	Medium	Medium	Medium
	Energy cost	Low to medium	Medium to high	Very low	Low	Low	Low	Low
	Cycle time present	High	—	High	Low	High	Medium	Medium
	Future	Medium	Medium	Medium	Very low	Low	Very low	Low

forming systems. It is important that design and manufacturing engineers utilize the correct forming processes, based on applicability of part shape, part size, material required, number of parts to be formed, tooling and other pertinent manufacturing considerations.

In order to select the correct forming process, it is necessary to determine the most applicable process for a specific part shape. In comparing the high-velocity and low velocity forming methods, it will be noted that deep recessing, shallow recessing and bulging operations encompass the greatest percentage of parts shapes applicable to the high velocity systems, the lesser percentage of applicable part shapes are flanging and plane contouring operations. Simple bending and linear contouring have little or no applicability.

Conventional forming methods are adequate for mass production of every part shape within the part size and energy capabilities of the respective machines. Although bulging operations are inherently difficult by conventional means, they are highly applicable to high-velocity forming due to the particular strain distribution requirements.

Advanced tooling and heating methods will extend the capabilities of conventional machines to larger part-sizes for forming materials with improved ductility at high temperatures (such as titanium) but large area parts will still require high tonnage equipment.

A major advantage of the high-velocity systems is the ability to form one-piece complex part shapes in a single operation, whereas conventional methods may require several operations and result in a welded structure. Preforms by conventional methods are frequently utilised in high velocity forming to reduce multiple-stage forming operations to a single forming and sizing operation.

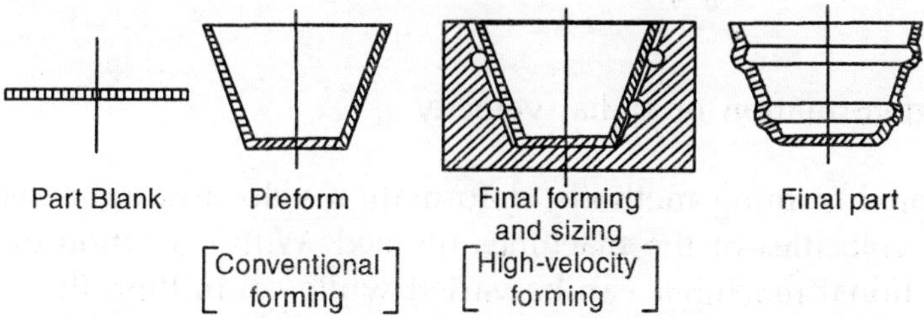

Fig. 14.1. *Conventional preforms for high-velocity forming*

DEFORMATION VELOCITY

In sheet metal forming, two velocities are generally considered. The first is deformation or radial velocity, defined as the time rate of change of displacement $\left(\frac{du}{dt}\right)$ in the direction of the applied force of any point on a part blank as it is formed to the final part shape. The second is the relative velocity V_R between two points on the surface of the part blank with an initial gage length (L_o). Consider the expanded tube in figure 14.2. The relationship of the strain rate (e°) and the relative velocity V_R of points A and B as the tube expands is given by

$$e^{\circ} = \frac{L_f - L_o}{L_o dt} = \frac{dL}{L_o dt} = \frac{V_R}{L_o}$$

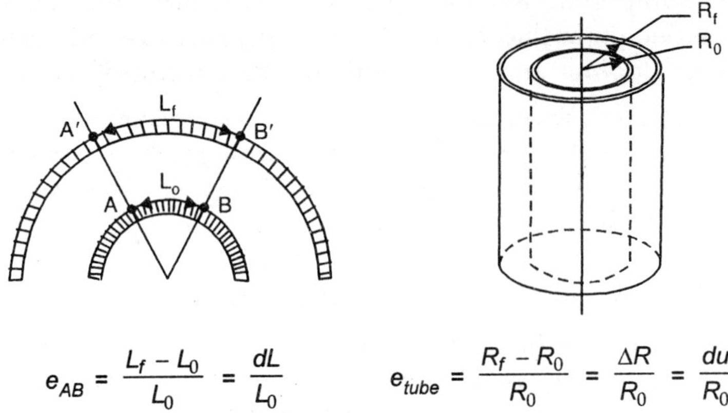

$$e_{AB} = \frac{L_f - L_0}{L_0} = \frac{dL}{L_0} \qquad e_{tube} = \frac{R_f - R_0}{R_0} = \frac{\Delta R}{R_0} = \frac{du}{R_0}$$

Fig. 14.2. *Circumferential strain for tube expansion*

The strain rate for the expanding tube is given by

$$e^{\circ} = \frac{R_f - R_0}{R_0 dt} = \frac{dR}{R_0 dt} = \frac{du}{R_0 dt}$$

Therefore, the relationship for the relative particle velocity and the deformation velocity is given by

$$V_R = \frac{L_0}{R_0} \frac{du}{dt}$$

where $\frac{du}{dt}$ is the deformation or radial velocity.

For conventional forming methods, deformation velocities are essentially the same as the closing (head) velocities of the machines utilized. With exception of the drop hammer, tonnage of conventional machines can be varied while controlling the closing velocity. For this reason, conventional machines are advantageous for forming strain-rate sensitive material. The drop hammer, depending on gravitational and pneumatic acceleration of a large fixed mass for deformation energy, is more closely associated with the high-velocity forming system although maximum velocity is somewhat less than that of the advanced pneumatic-mechanical machines.

Deformation velocity for high-velocity forming systems are influenced by the type of energy and the part geometry. High explosive and capacitor discharge systems utilize shock-wave action where the total energy of the system is dissipated in microseconds, whereas chemical deflagration processes utilize volumetric expansion of the gas bubble where the time period is milliseconds, lower velocities are achieved by the latter type.

Due to positioning of the explosive charge or electrodes, high-explosive and capacitor-discharge processes are particularly influenced by part geometry.

MATERIAL BEHAVIOUR

To adequately compare conventional and high-velocity forming conditions, it is helpful to consider the following

(1) effect of velocity on ductility and material strength

(2) effect of relative velocity of the part blank and die on friction

(3) geometric stability

(4) wave effects.

For dynamic loading conditions, ductility has been found to be either greater or less than ductility under static condition, depending on the material. Although increased ductility is desired, the primary objective is to satisfactorily form a given part shape and material without exceeding the initial velocity. Regardless of ductility response, material yield strength is increased for dynamic loading conditions. Due to increased yield strength, materials with increased ductility for dynamic loading conditions will require higher deformation energy for high-velocity forming than for static conventional forming.

Friction at the part blank and die interface is a major consideration in sheet metal forming. The coefficient of friction generally decreases with increasing relative velocity between the part blank and die, but increased velocity results in increased temperature and lubrication breakdown. However, extremely high velocity may generate sufficient temperature to melt a thin layer of metal to the blank and die interface to act as a lubricant.

When a shock wave is initiated by the source and transmitted through a medium less dense than the part blank, the shock wave is partly transmitted by the blank and partly reflected back as a compressive shock wave.

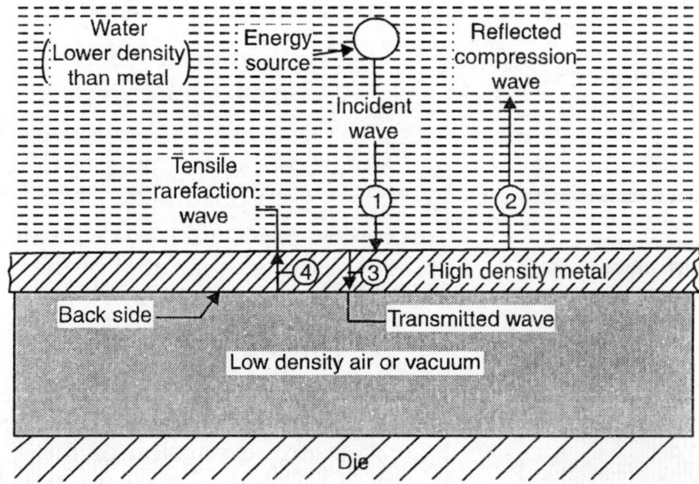

Note : water to metal ⟶ low to high impedance match
metal to air ⟶ high to low impedance match

Fig. 14.3. *Wave sequence for shock-wave forming processes*

The reflected wave (2) is compressive due to the low to high impedance match of the fluid medium (water) and the higher density metal, the compressive wave is the reactive force of the incident wave (1) and causes the metal to deform towards die. Microsecond later, the transmitted incident wave (3) reaches the back side of the blank and is reflected back as a tensile rarefaction wave (4) due to the high-to-low impedance match of the shock wave travelling from high-density metal to low-density air. The reflected tension wave is partly transmitted back into the water, causing the water to cavitate.

STRAIN DISTRIBUTION

For comparison of strain distributions for various loading conditions, it would be advantageous to select one process and part shape common to both conventional and high-velocity forming. A typical example is deep recessing of spherical segments (dome ends) from flat discs. Deep recessing is most commonly accomplished with the following conventional machines : hydraulic press, rubber bladder press etc.

The figure 14.4 illustrates the three basic types of strain distribution to be considered. Only radial strain will be considered for the three basic types, since the radial strain is greater than the circumferential strain and is the limiting strain to fracture. In each case, it is assumed that sufficient restraint is applied to the part blank for a no-draw condition.

Type I, is the strain distribution curve for deep recessing with a punch arrangement such as forming with the conventional sheet stretch press or the hydraulic press. An uniform straining occurs at the centre of the part as the material initially deforms around the punch. Such straining continues until sufficient deformation is accomplished to result in a localised static condition, due to friction, at the interface of the punch and blank. As a result of this condition, stretching of a circumferential band of material in the unsupported region begins; diffuse necking and ultimate strain occur in this region of instability.

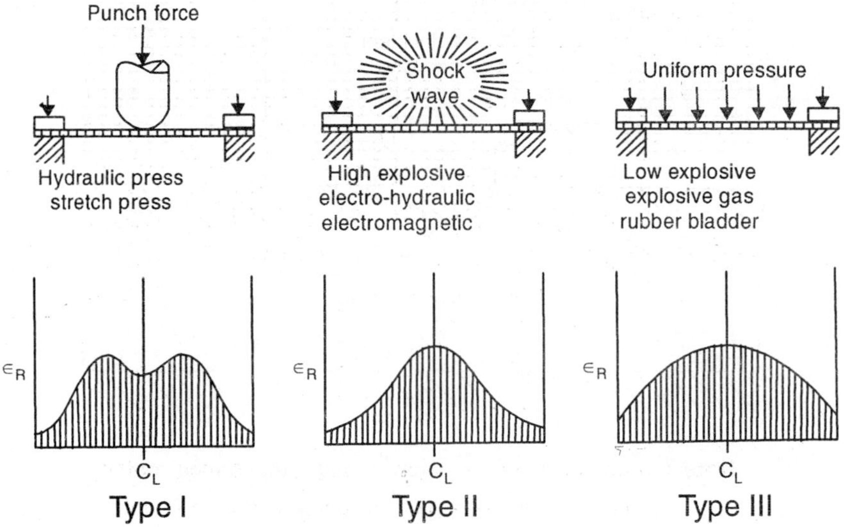

Fig. 14.4. *Three basic types of strain distribution*

In type II, the flat disc, when impulsively loaded, deforms vertically as a flat plate, while the inclined sides remain relatively motion-less, the part assumes a conical shape with little work being done in the circumferential direction. As a result, the maximum strains occurs at the apex of the cone.

Type III. The processes exert uniform pressure on the part blank during the entire period of deformation. The curve is not influenced by frictional effects or shock wave

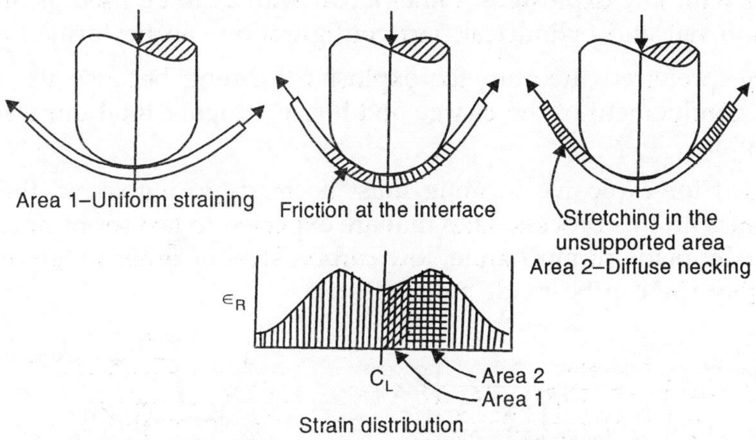

Fig. 14.5. *Effects of recessing with a punch arrangement*

14.3 EXPLOSIVE FORMING PROCESS

Explosive forming is a process that forms parts using the explosive force of a chemical reaction. The process uses energy in the form of a high-velocity shock wave to deform workpieces at rates of several hundred meters per second. The highest forming pressures of any HERF process are obtained with explosive forming; several thousand mega pascals.

PROCESS PRINCIPLES

There are two types of explosives which are used in explosive forming; low and high explosives. Low explosives such as explosive gases and gun cartridges, are commonly used only for light-duty forming applications. The low-explosive forming process is usually performed in a confining die which completely encloses the workpiece. Confinement maximizes the explosion energy, which is transferred to the workpiece. Figure 14.6 depicts a confined cartridge-type system. Although confined low explosive forming is useful for high precision applications, the relatively low forming pressure of less than 700 MPa limits its use to thin-walled workpiece.

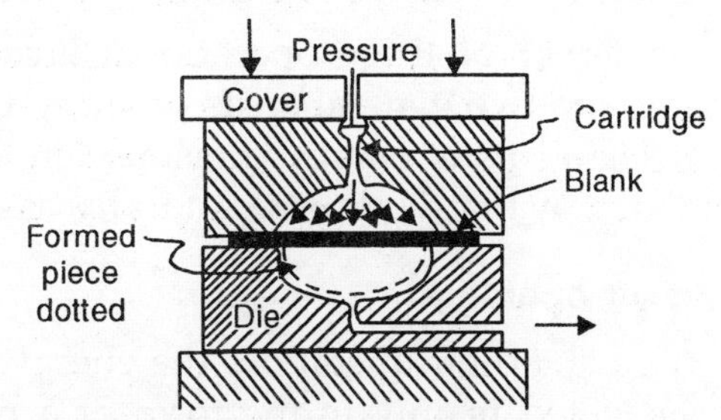

Fig. 14.6. *Confined HERF operation using a low-explosive cartridge-type energy source*

The high-explosive forming process generates pressures exceeding several thousand megapascals in a time span of approximately 100 μ sec. A detonating fuse, such as primacord, is the most commonly used explosive for forming because it can be easily formed into shaped charges. Large workpieces often require other explosives such as TNT, RDX, PETN, 60% dynamite etc.

A shockwave sent off by high explosives is so great that the explosion does not need to be confined as with low explosives. Either air or water can be used as an energy transfer medium; and both flat and cylindrical part configuration can be formed.

Water is the preferred medium for explosive forming because its incompressibility provides greater confinement of the charge and hence a higher total impulse efficiency than that provided by air.

The dies used for explosive forming must be made to withstand the extremely high pressures inherent with this process. Dies that are expected to last for hundreds or thousands of cycles are usually made of mechanite, low carbon steel or other tough, ductile metals. In expensive, one-shot die can be made from plaster.

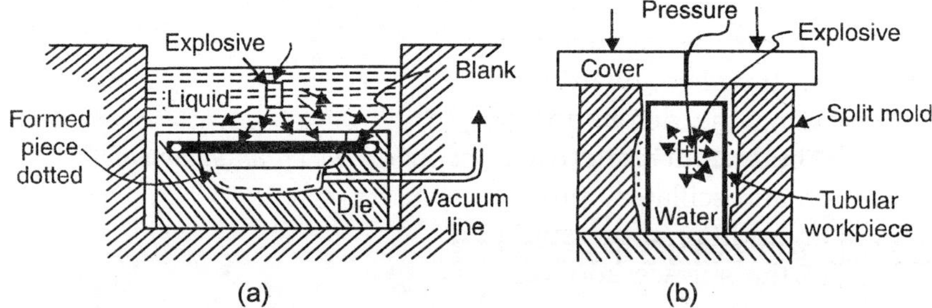

Fig. 14.7. *High-explosive forming operations : (a) forming flat workpiece; (b) forming cylindrical workpiece*

TYPES OF EXPLOSIVE OPERATIONS

In the explosive working of metals, large differences in the energy requirements and in the resulting behaviour of the workpiece will occur, depending on whether a stand-off operation or a contact operation is performed.

In either of these operational areas, the behaviour of the system can normally be divided into two general areas of study. One area treats the energetics of energy release and the mechanisms of energy transmission to the workpiece. The other considers the response behaviour of the workpiece and any associated components.

Stand-off operation

In a stand-off operation, the energy is released some distance from the workpiece and is propagated mainly in the form of a pressure pulse through some intervening medium. Peak pressure at the workpiece range from several thousands to several hundred thousands pounds per square inch, with most operations performed in the lower pressure area. The working times are normally measured in milliseconds and the metals displacement velocities in terms of several hundred feet per second. The motion and subsequent deformation of the workpiece are primarily dependent on the external forces with little, if any, energy appearing in the metal in the form of a stress pulse.

Contact operation

In contact operation, the energy is released while the energy source, usually an explosive charge, is in intimate contact with the workpiece. Interface pressures acting on the surface of the metal may be several million psi, with associated working times measured in microseconds. Pressures, times and impulse values are directly related to the metal-explosive combination and to the geometry of the metal-explosive system. In most contact operations, high-intensity, transient stress waves are induced in the workpiece and then propagate through the metal and displace, deform and possibly fracture it.

ENERGY DATA FOR UNDERWATER STANDOFF OPERATIONS

In an underwater standoff operation where an energy source, say an explosive charge, releases its energy some distance from the workpiece, it is necessary to know something about the mechanisms of transmitting that energy through the water, and the quantitative values of pressures, impulses, times, which exist at some distance from the energy source. Data of this type come from relationships determined by a large number of underwater research programs.

With the release of explosive energy, a high intensity, spherically expanding pressure pulse is initiated which travels outward through the water with a velocity of about 4800 fps in fresh water. Following the detonation of the explosive, a gaseous bubble is produced which in the general behaviour pattern expands to an initial maximum radius, collapses, and then goes through an oscillatory process as it rises or migrates through the water. Each oscillation produces an additional pressure pulse, but of a much lower intensity. The bubble will eventually rise to the surface and vent to the atmosphere.

The pressure pulse

The pressure pulse generated by a detonating charge will propagate outward through the water with an expanding spherical front, constantly reducing in peak pressure and unit impulse due to energy loses to the water and the divergent nature of the pulse. At any given distance from the centre of the charge, the pulse will have a pressure time profile of the type shown below. This can be approximately represented by the expression

$$P = P_m e^{-t/\theta}$$

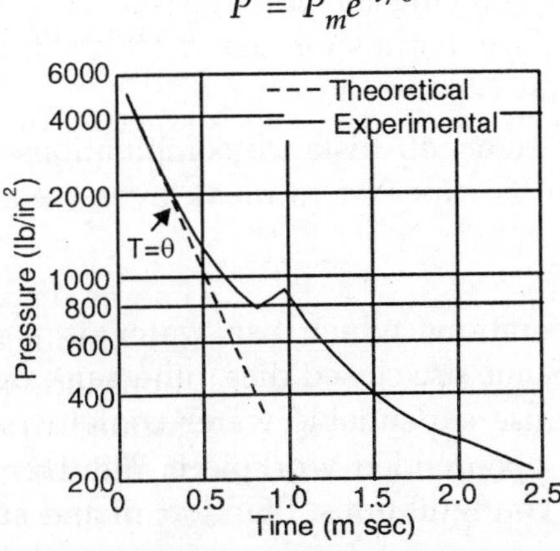

Fig. 14.8. *Generalized pressure-time profile of pulse from underwater explosion*

where P is the pressure as a function of time

P_m is the peak pressure at that distance

t is the time after arrival of the pressure front

θ is the time constant characteristic of the charge weight, type of explosive, and distance from the charge, and represents the time that it takes the pressure to fall to $1/e$ of its peak value.

The gas bubble

Following the detonation of the explosive, a high pressure gaseous bubble is produced which goes through an oscillatory motion until it vents at the surface. With each oscillation, energy is released to the surrounding water. Measured data for large charges reported by Cole indicated that the pressure pulse accounted for about 60 percent of the energy available, the first oscillation for 25 percent, and the remaining behaviour of the bubble for the other 15 per cent. These values will probably vary with the charge size and type of explosive.

The maximum bubble radius occurs during the first period, the size of the bubble being smaller with each subsequent oscillation. A general expression for determining maximum radius of the bubble is

$$r_m = J\left(\frac{W}{dt33}\right)^{1/3}$$

where r_m = radius in feet

W = weight of charge in pounds

d = depth in feet to which the charge is submerged

J = an explosive constant

During the oscillatory process, the bubble will migrate normally rising to the surface and releasing its remaining energy to the atmosphere and producing a water plume. If the charge depth is small compared to the bubble radius, the bubble may vent during the first period. In other instances, the bubble may be attracted to nearby rigid underwater surfaces, breaking over the object and allowing the compressed gases to do additional work on the object before rising to the surface. For a short stand off distance, the bubble may break over the workpiece during the first expansion.

The charge weight and stand off distance combinations for many forming operations are such that the bubble can be expected to break over the surface of the work piece.

System types

Explosive stand off operations which use water as a medium are performed in a variety of different systems. Some use closed dies, others use open dies. Many are performed in large water tanks, others use expendable water containers, and still others contain the charge and water within an open ended workpiece. The charges come in a large range of weights and in a variety of configurations. Standoff distances range from a few inches to many feet.

ENERGY DATA FOR EXPLOSIVE CONTACT OPERATIONS

Representative pressure data

When an explosive charge is detonated in contact with a metal surface, a detonation front is initiated which proceeds through the explosive at a constant, reproducible velocity. Detonation velocities for the common industrial and military explosives are well established and can be obtained from many sources. When in doubt, this velocity can be obtained by means of a high speed streak camera.

A high pressure is induced in the metal at the contact surface and a high-intensity transient stress pulse is initiated. The magnitude of the induced pressure depends on the metal-explosive combination and on the orientation at which the detonation front impinges on the metal surface. For proper design of a contact operation determination of pressure magnitude is required. The determination of pressure magnitude induced in the metal by normal incidence detonation can done by an equation such as

$$P_m = \frac{2P_x(\rho_m C_l)}{(\rho_m C_l + \rho_x D_x)}$$

where P_m is the pressure induced in the metal

P_x is the detonation pressure of the explosive

ρ_m is the initial density of the metal

ρ_x is the initial density of the explosive

C_l is the sound velocity of the metal

and D_x is the detonation velocity of the explosive

If the detonation pressure of the explosive is not known, an approximate value can be obtained from the following working relation.

$$P_x = \frac{\rho_x D_x^2}{4}$$

This equation is based on the assumption that the density of the explosion products is 30 percent greater than the initial density of the explosive.

Explosive	ρ_x (gm/cc)	D_x ft/sec
50/50 Amatol	1.55	21,100
Composition B	1.66	25,600
Composition C-3	1.60	25,000
50/50 Pentolite	1.65	24,400
Tetryl	1.71	25,700
TNT	1.56	22,600

Representative Explosive Values

The two extreme cases by which system orientation can affect the induced pressures are shown in figure 14.9.

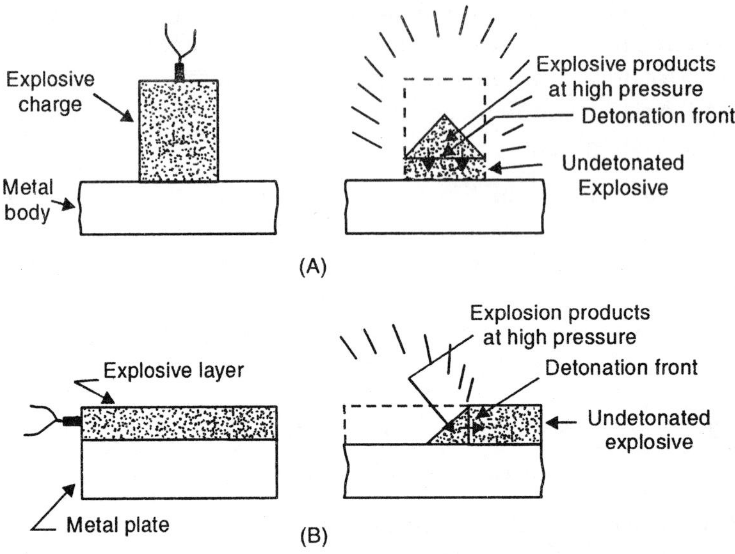

Fig. 14.9. *Normal and oblique incidence of detonation fronts. (A) normal incidence; (B) oblique 90° incidence*

At A, the detonation front is striking normal to the metal surface (normal incidence); in B, the detonation front is moving in a direction parallel to the metal surface.

Table 14.2 compares values of pressures produced in several metals by contact charges of composition B for above two extreme orientations.

Table 14.2. Comparison of Pressures Induced by Normal and Obliquely Incident Detonation Fronts using Composition B Explosive

Material	Normal incidence	Oblique incidence	Pressure ratio Pn/Po
Aluminium	5,200,000	2,800,000	1.85
Copper	7,000,000	3,000,000	2.33
Iron	6,800,000	2,900,000	2.34

EXPLOSIVE FORMING TECHNIQUE FOR STANDOFF OPERATION

The Process. In this process, a high explosive is detonated, and its energy is transmitted through a fluid medium, commonly water, to a workpiece which is held, typically, on a female die.

The space in the die behind the workpiece is commonly evacuated to avoid adiabatic compression and heating of the entraped air. If not controlled, this could result in burning of the rear face of the workpiece and the die.

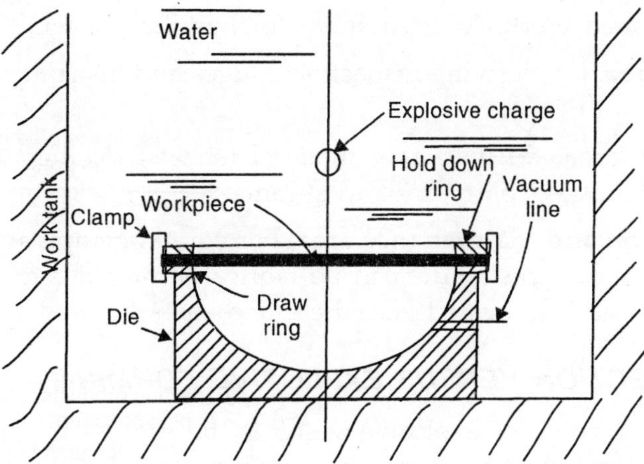

Fig. 14.10. *Schematic view of typical explosive forming operation*

The energy transfer medium need not be water. Hot forming calls for molten salts, molten metals, sand, oil or other media which can sustain the desired forming temp. Air can also be used as an energy transmission medium. Many other media are possible, but for the cold fabrication of large workpieces, cost and convenience considerations leave air and water as the most suitable choices, with water being superior because of its reduced compressibility and shock-impedance matching characteristics.

Energy source

The underwater detonation of a high explosive yields a shock wave and bubble of hot gaseous detonation products. The shock wave is normally the major energy source in this process, although secondary bubble phenomena can become important under certain conditions. Generally, these explosions occur at relatively shallow depths, sometimes so shallows that the bubble breaks through the surface of the water. The secondary shock-wave reflection effects from the water surface and from the tank wall somewhat modify the simple analysis. In the case of the water-air interface, shock reflection causes an inward moving rarefaction which reduces the impulse delivered to the workpiece if water surface is relatively close to the charge.

In general, any homogeneous explosive of a reasonable sensitivity is suitable as an energy source. Dynamites and liquid explosives have been used.

Facility Requirements

For an explosive forming facility, the general requirements are as follows :

1. **Operations area.** (*a*) either a natural lake or a river, an artificial tank in the ground (*b*) water supply and pump if not naturally available (*c*) vacuum force-pump and pressure gauge (*d*) portable electrical power supply.
2. **Explosive handling facility.** (*a*) storage magazine (*b*) preparation area (*c*) shelter and control barricade.

3. **Material handling.** (*a*) Crane or truck with 'A' frame. (*b*) Trollies for transporting dies and finished work. (*c*) Equipment for handling sheet.
4. **Inspection area.** (*a*) receiving inspection gauges and instruments (*b*) final inspection (*c*) reject or rework storage.
5. **Inert storage area.** (*a*) inert raw material (sheets) (*b*) inert components for H.E. operation (*c*) finished parts prior to shipment.
6. **Die fabrication and maintenance area.** For very large operations or where dies are of concrete, ice or kirksite, die can be fabricated on the site. For other cases, dies are purchased and fabricated elsewhere and then shipped to the forming site.

EXPLOSIVE CORRECTION VERSUS EXPLOSIVE FORMING

The explosive correction, or explosive sizing is the process in which either an out-of-tolerance part or a preformed part is explosively deformed to its desired final configuration. Explosive correction is a special case of explosive forming.

Explosive correction was most appropriate process for salvaging parts which did not meet the tolerance requirements after fabrication by conventional methods. There are a number of rocket hardware components which can suffer from such troubles. A common one involves the deviation from desired diameter and contour in the dome or the rocket engine closure. A poor match between dome and cylinder causes serious assembly problems. Another common trouble involves the distortion of the dome after bosses are welded on. Both problems have been satisfactorily handled with explosive correction techniques.

Explosive correction process is applicable not only to the pressure vessel closures but also to the complete pressure vessel assemblies as well. Fabrication from preforms is also essentially an explosive correction process.

Explosive forming from flat sheet is a more complex process than explosive correction. However, despite its greater difficulty, explosive forming from flat sheet is overall a more economical process for fabricating closures than the hybrid process involving conventional drawing followed by explosive correction.

Explosive Compaction

In 1956 effect of explosive pressure on physical and chemical properties of various materials were observed. A number of interesting phenomena, including chemical composition was noticed. Numerous investigators have become interested in the problems of explosive compaction. A very interesting application is concerned with the explosive compaction of refractory metal powders,

The following advantages have been observed :

(*i*) Extremely high pressures can be applied.

(*ii*) The area of the object being pressed is not limited by the maximum total force.

(*iii*) Very high compact densities can be obtained.

(*iv*) Shrinkages on sintering can be markedly reduced.

14.4 GAS MIXTURE PROCESS

High velocity forming, with combustible gas mixtures, employs the energy released by the chemical reactions of a fuel with an oxidizer. The energy released by the reaction may be applied directly to the workpiece, through a transfer medium, or through a mechanical device such as a press.

CHARACTERISTICS OF GAS MIXTURE FORMING

Gas mixture forming is normally performed in an enclosed container, it requires more extensive tooling and greater safety considerations in tool design. However, it is particularly suitable for rapid cycling and forming of small tubular parts, complex shapes and large thin sections. This suitability is derived from the following characteristics of a gas mixture :

(1) A gas mixture can be placed in position, ready to fire, in a short time with no requirements for rigging of charge.

(2) A gas mixture assumes the shape of its container and, therefore, approaches the optimum shape regardless of the quantity used. This removes the variable of charge shape and stand off distance. The shaping is important in providing a uniform force over the entire surface of a complex part.

(3) The gas mixture can be easily adjusted within flammability limits, by varying the amount of fuel and oxidizing agent, the amount of diluent gas, and the initial pressure of the charge. This provides a versatile production setup.

(4) The gas mixture can be detonated like a high explosive, but it is normally used in the adiabatic combustion range. The adiabatic combustion forces are applied at slower rate than those emanating from a detonation. This is an important factor in forming large, thin sections. Detonation might be used on small parts where forming forces might be high, and adequate strength can be practically provided in the tooling.

THE GAS-MIXTURE SYSTEM

The direct-contact system for gas mixtures requires a system designed specifically for handling gases. The basic gas mixture system contains :

(1) separate sources of pressurised fuel, oxidizer and diluent gases.

(2) mixing and combustion chamber of adequate strength to contain the maximum pressures to be generated.

(3) a female die of the desired configuration into which the workpiece is to be formed.

(4) a gas transfer system consisting of suitable valves, regulators and lines to provide complete control of the transmission and measurement of each gas independently. This arrangement must assure separation of the gases until they have been injected into the combustion chamber.

(5) an ignition system such as a glow wire or spark plug.

(6) a clamping, sealing and vacuum system to allow removal of air from between the workpiece and the die.

(7) an exhaust valve system to allow removal of fired gas pressure prior to disassembly of the tool and removal of the workpiece.

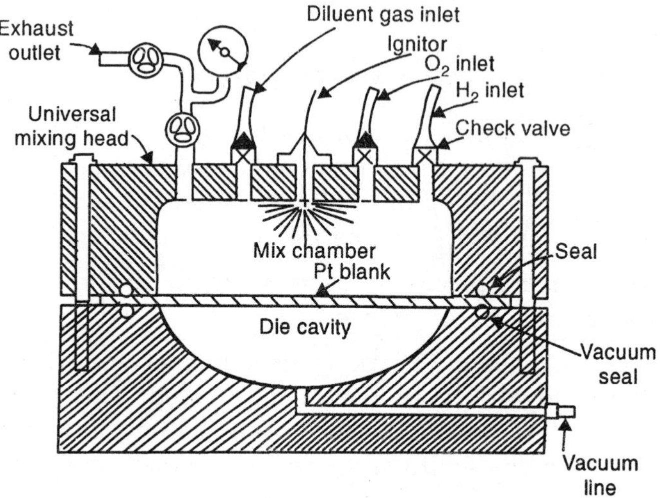

Fig. 14.11. *Schematic of universal mix chamber*

The features listed above are essential to the basic safety and operation of a gas-mixture system. Other features, such as premix chambers, rapid charge setups and automated systems for die operation and part handling may be incorporated to improve the speed of operation.

THEORY OF THE COMBUSTIBLE-GAS-MIXTURE REACTION

When a mixture of combustible gases is confined within a closed vessel and ignited, the reaction may occur in three phases :

1. **Adiabatic combustion.** The velocity of the flame front is subsonic and the changes in pressure, temp. and volume follow the laws of an adiabatic reaction.
2. **Unstable or transient detonation.** An extremely turbulent reaction is accompanied by severe shocks and over-pressures. This is a transient phase and is usually followed by stable detonation.
3. **Stable detonation.** The reaction occurs at a fixed supersonic velocity within a thin segment of the flame front, and with the unburned gas ahead of the shock front at rest.

The following thermochemical reactions for hydrogen and oxygen indicate what happens in the adiabatic combustion reaction :

$$H_2 + \frac{1}{2}O_2 \rightleftharpoons H_2O + 57{,}107 \text{ calories/gm mol}$$

$$OH + \frac{1}{2}H_2 \rightleftharpoons H_2O + 67{,}107 \text{ calories/gm mol}$$

$$2H \rightleftharpoons H_2 + 103240 \text{ calories/gm mol}$$

$$2O \rightleftharpoons O_2 + 117172 \text{ calories/gm mol}$$

Le Chatelier's principle on equilibrium mixtures assists in an understanding of the end result. This principle may be summarised as follows :

1. Increase in temperature favours action that uses heat.
2. Increase in pressure favours actions that reduces volume of the mixture.

Both temp. and pressure increase in an adiabatic combustion in a closed vessel. Since temp. increases drive the reactions to the left while the pressure increases drive them to the right, an intermediate point of equilibrium is reached. Consequently, the end products are not only H_2O, but H_2, O_2, H_2O, OH, H and O.

In a closed vessel, the flame will normally propagate itself in a spherical flame front from the point of ignition until it is quenched upon the vessel wall.

In adiabatic combustion, as the first increment of gas burns and expands, the unburned gas is compressed slightly. The next increment of burning further compresses the unburned gas and partially recompresses the previously expanded and burned gas. This process continues in the spherical flame front until quenching occurs.

The flame propagation phenomena also involves velocity. Until it reaches sonic velocities, the propagation of flame is undergoing a continual acceleration. The acceleration force is a function of the atomic attraction forces of the particular atoms involved, their separation distances and the degree of agitation. The velocity at any time, therefore, will depend upon the accelerating force and the distance travelled.

SELECTION AND CHARGING OF THE GAS MIXTURE

The selection of a fuel, oxidising agent, diluent, mixture ratio, and initial charge pressure is subject to the following over lapping considerations :

1. **Safety.** (*a*) Unstable detonation and resulting localized high over pressures should be a matter of planned usage and not an unexpected occurrence. (*b*) Gases or their combustion products should not be extremely toxic.
2. **Forming.** (*a*) Gas mixture and initial charge pressure must be capable of producing required forming forces. (*b*) Rate of combustion must be within mass limitations of thin wall workpieces. (*c*) Detonation and the chance of localised high over-pressures on the workpiece will normally be avoided.
3. **Handling.** (*a*) Gases should be "permanent gases", that is, they should remain in a gaseous state at charging pressures and temperatures to allow accurate determination of charge amounts and ratios. (*b*) Gases or combustion products should not be extremely toxic or corrosive. (*c*) Gases should not require use of exotic handling and transmission equipment.

4. **Combustion.** Gases should be permanent gases at charging pressure and temperature to assure thorough mixing and uniform combustion.
5. **Metallurgical.** Gases or their combustion products should not be metallurgically detrimental to either the workpiece or the tool.
6. **Cost.** Fuel gas, oxidising gas, and diluent gas should be within practical cost limitations while satisfying other requirement.
7. **Design.** Mixture and initial charge pressure should not produce peak pressures beyond those selected for tool design criteria.

The following list covers common gaseous fuels, oxidizers, and diluents with an analysis of their capability to meet requirements :

1. **Fuels.** **(*a*) Hydrogen.** Meets all requirements; has a great amount of documented experimental data; is readily available in high-pressure containers. **(*b*) Ethane and methane.** Meet all requirements, although some toxic carbon monoxide can be produced as a product combustion. **(*c*) Natural gas.** generally meets all requirements; is lowest in cost; not commonly available at high pressure; can produce some carbon monoxide as a product of combustion; requires analysis for hydrocarbon composition if analytical studies are desired. **(*d*)** Butane, propane, acetylene and carbon monoxide are not satisfactory as fuels since they either liquefy, are toxic, or undergo secondary decompositions.
2. **Oxidizers.** **(*a*) Oxygen.** Meets all requirements within reason; readily available in high pressure containers. **(*b*) Air.** Meets requirements, cheapest and most readily available; may require drying and filtering; may not contain adequate oxygen for desired percent diluent.
3. **Diluent.** **(*a*) Helium.** Meets all requirements but has high cost, its low heat capacity and light weight allow rapid reaction and high peak pressure, but require high dilution to prevent detonation used primarily for small systems where inert characteristics justify cost. **(*b*) Argon.** Some chemical and thermodynamic properties as helium, however, greater mole weight slows reaction velocity and leads to poor mixing which promotes unstable detonation. **(*c*) Nitrogen.** Most practical and cheapest diluent both in pressurised containers or as a constituents of air, intermediate heat capacity and nearly same mole weight as oxygen assure uniform mixing and control of reaction. **(*d*) Carbon dioxide.** Generally meets all requirements of a diluent, readily available in pressurised containers at low cost, provides greatest damping effect.

TOOLING AND GAS SYSTEM DESIGN REQUIREMENT

The gas system must be designed and constructed to allow complete control of the flow of each gas into the combustion chamber. In addition to on-off valves and pressure regulators, each line should have check valve. The chamber itself must be completely sealed so that no leaks may develop. It must fulfil its functional purpose and guarantee structural integrity. The later may encounter a practical limit in large tools, since the over pressure experienced in unstable detonations can range up to 50 times initial pressure.

Where die design is not adequate to contain detonation pressures, the utmost personnel safety precautions are mandatory until a sufficient background of test experience is available with the particular tool, gas mix and initial pressure.

The chamber geometry should avoid sharp projections, or pockets, which can cause incomplete mixing and unstable reactions. Small injection ports are desirable for strength, mixing and to quench back firing into the gas lines, should a valve malfunction. In addition, the chamber material should have good thermal conductive properties. The excess heat of the reaction must be rapidly transferred from the chamber after the forming operation has been completed. Another chamber design requirement is an exhaust system to relieve the chamber of residual pressures after the reaction has gone to completion.

The spark ignition systems will not reliably initiate combustion above five atmospheres initial pressure. A 28-volt aircraft coil and turbine plug which incorporates spark and glow features is an improvement, but it too becomes unreliable above six to seven atmospheres. An electrically heated glow wire which extends into the chamber has proven to be the most reliable system at the pressures normally encountered. This device requires a periodic replacement of the glow wire. Electric squibs (explosive caps) are effective in initiating reaction, but the explosive shock wave from the cap may initiate a detonation of the gas mixture. Another important consideration in designing the ignition system is the location and number of ignition points. Normally, one point will suffice, but if the size of the chamber raises questions about critical path lengths for the flame front, the path length may be reduced by proper location of several igniters to initiate combustion at a number of points simultaneously.

Sealing of the chamber to the workpiece presents few problems. The use of good thermal conduction principles usually will prevent damage to the sealing material. Seals of synthetic rubber compounds have proven satisfactory. The design of the gas-mixture system must be such that the system will not be easily contaminated by hydrocarbon lubricants. These compounds are fuels, and they may result in a premature reaction prior to completion of mixture charging on the planned system ignition.

APPLICATIONS OF GAS MIXTURE PROCESS

Gas mixtures are best suited to complex shapes, large sections of thin gage materials and operations where rapid automated cycles are required. The infinite variations of the mixture proportions that are possible, provide an experimental versatility in adjusting the charge to suit the forming requirements.

14.5 ELECTRO-HYDRAULIC FORMING

Electro-hydraulic forming (also known as electroshape or electrospark forming) is essentially the same in principle as explosive forming. The primary difference is that electrohydraulic forming shapes parts using the pressure from a spark-generated shock wave versus one generated by a chemical explosion.

A capacitor bank stores electricity until a switch closes a circuit to the water filled forming tank. When energized, an aluminium or magnesium bridge wire placed between electrodes is instantly vaporised and provides a plasma channel in the water for a spark to cross. The spark generates a shock wave that propagates radially, forcing the part into the die. An alternative to using the bridge wire is to increase the capacitor voltage from 4500 to 20000V. The higher voltage is able to generate a spark without the help of a bridge wire and is more convenient because there is no need to replace a wire after every part.

Electrohydraulic forming is often preferred over explosive forming because of its shorter cycle times and better potential for safe operation. Pressures of 1200 MPa can be obtained with pulse energies ranging from 6 to 200 Kilo joules. The liquid medium used in the process must be a incompressible as possible : water, glycerin, or light oil are commonly used materials.

THEORY OF THE METHOD

Electrical aspects. Electrical energy in a charged capacitor is expressed as

$$U = \frac{1}{2}CV^2$$

where U = Energy in watt-seconds or joules

C = Capacitance in farads

V = Charged voltage

Since the voltage occurs as a squared term, the energy varies as the square of the voltage. This conclusion might lead to the assumption that when higher energy is desired, a higher voltage should be used. But the physical size of a capacitor is dependent mainly upon its energy storage capability, not its voltage or capacitance alone. Actually, a somewhat larger container is required for a capacitor to store a specified amount of energy at 20KV than at 5KV, and a still larger container is required to store the same amount of energy at 50KV.

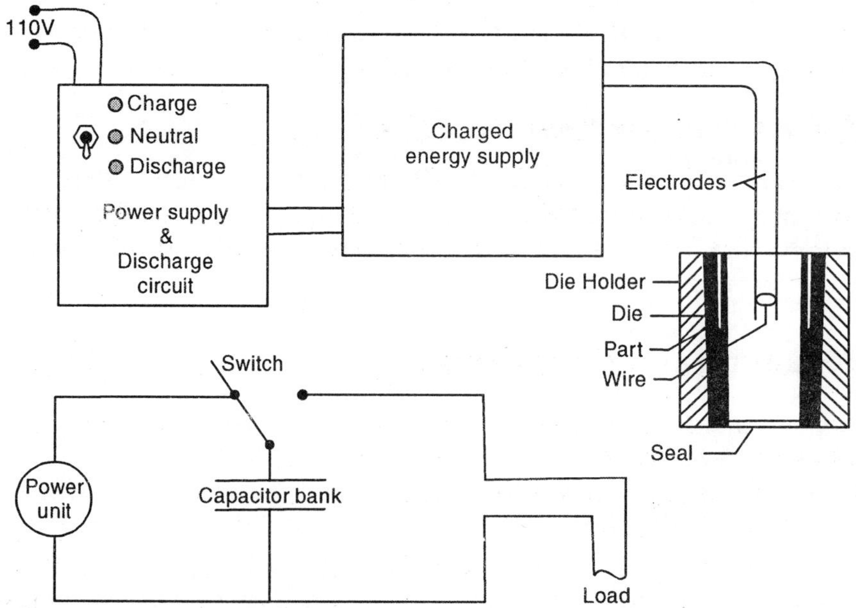

Fig. 14.12. *Block diagram and schematic of electro-hydraulic forming system*

The capacitor discharge units for electro-hydraulic forming developed during the late 1950 and early 1960's had energy storage capabilities of 6000 to 40000 joules. Either 230 or 115 volts is suitable for the basic power source for such energy levels. Even though the conversion to shock of the spark discharge or exploding wire is very complex, the operation of the mechanism is very simple. The capacitor bank is charged by means of a power supply. The stored energy is then dumped into the spark gap, or the wire to be exploded, by some extremely fast-operating triggering device.

Conversion into shock

To clarify the various physical processes which follow one another, calculations were made for the exploding of a medium-size (0.04 in dia) aluminium wire. The current of just a few thousand amperes will heat a wire to almost 10000°F in 0.1 microsecond. Inertial forces, assisted at first by magnetic fields, contain the material. The pressure on the wire's axis increases to many thousand atmospheres. This leads to the explosive expansion. At the beginning of the expansion, the material becomes nonconducting compressed gas and the current is interrupted. The expanding metallic vapour pushes a shock before it. In a few microseconds, the gas has thinned out sufficiently to permit the initiation of a normal gas discharge.

The average wire temp. can be determined from the energy input and the mass of the wire if thermodynamic equilibrium exists. In the solid and liquid state, the thermodynamic equilibrium probably is reached in short time intervals, compared to a millimicrosecond since electron-relaxation times are typically of the order of 10^{-13} sec. At ten millimicroseconds, enough energy has been put into the wire to melt it completely. At 12 millimicroseconds, the wire begins to vaporize at one atmosphere and at 17 millimicroseconds, enough energy has been developed to completely vaporize it.

Large magnetic pressures, caused by the large current flow, raise the temp. sufficiently for vaporization to occur from a minimum at the wire's surface to a peak along its axis. This vaporization probably occurs preferentially from the wire surface. Initially, the surface begins to vaporize into the surrounding medium at a little more than one atmosphere, but it appears that vaporization occurs so rapidly that the outer vapour layer does not move an appreciable distance before the next underlying layer is vaporized. Consequently, huge pressures are built up, resulting in progressively higher vaporization temp.

Electrical requirements

Power Supply Requirements. Power supply requirements depend upon the voltage rating of the capacitor bank, voltage of the power source, and the desired charging time. Major components of the power supply are a step-up transformer and the rectifier. The voltage rating of the power supply must be selected on the basis of the operating voltage rating of the capacitors. The current rating must be sufficient to permit charging the capacitor bank with the desired time limit. Charging times of from 6 to 30 seconds have been found to be practical. Shorter times would be necessary for a high production machine.

Storage System. The storage system consists of a suitable bank of capacitors connected together with either heavy conductors or bus bars so as not to create detrimental resistances.

One side of the capacitor bank must be connected through a triggering device to the gap. The stored energy must be delivered by the bus bars and rapidly dumped by the triggering device into the work gap by way of transmission lines sufficiently large to impose only very small resistance.

The size of the storage system will depend on the amount of energy desired. The amount of useful work obtainable from the conversion of electrical to mechanical energy has been found to vary from 10 to 40 per cent of the released energy, thus to accomplish 2000 ft-lb of useful work with one energy release, a bank having from 7000 to 26000 joules is required.

Voltage rating

If the spark discharge technique is to be used, voltage rating of the energy storage bank should be from 20000 to 30000 volts. This is necessary in order to obtain a longer gap within which the energy is to be discharged.

When using a initiating wire, a very efficient energy supply can be built from capacitors of 4000 to 7000 volt rating.

Other system parameters

In addition to energy level, there are other important parameters such as power supply and switching arrangement discharge voltage, circuit inductance, wire material, wire size and gap length. Wire material was also found to be a parameter, magnesium and aluminium have both been found to be very effective. Aluminium foil also is a good initiating material.

Force of the converted energy

In research upon the magnitudes of forces developed by exploding wires submerged in water, capacitor banks of four different capacitance ratings were investigated. A series of discharges at varying voltages were carried out with 120, 600, 1200 and 1800-microfarad capacitance. All the discharges were made using aluminium wire of 0.045-in-dia across a gap of $\frac{1}{2}$ inc. The 'standoff' was maintained constant at one inch.

From the data obtained, it was estimated that an average pressure of approximately 35000 psi acts at a distance of one inch from the 0.045-in-dia aluminium wire when 18000 joules are discharged into the wire.

Force required

The force required to form a given part depends upon many variables such as the size of the part, the kind of material being formed, the material thickness, and the amount of restraint imposed on the part by the tooling. It is worthwhile to point out the usefulness of the hoop tension equation

$$S = \frac{Pr}{t} \quad \text{or} \quad \boxed{P = \frac{S \cdot t}{r}}$$

S = yield strength of material, *psi,*

r = initial radius of part, in,

t = thickness of materials, in,

P = pressure, *psi*,

The above equation gives the minimum pressure required to produce a permanent set in a tubular part of a certain diameter, material and wall thickness. In order to produce significant deformation in a part, considerably more initial pressure must be produced. If the energy release is sufficiently large to cause the strain rate to exceed the critical impact velocity of the material being formed, fracture will occur from the over-shoot.

APPLICATIONS OF ELECTRO-HYDRAULIC FORMING

Bulging, forming, beading, drawing, blanking, and piercing can be accomplished. All of these operations are performed by conventional sheet-metal working equipment such as rolls and presses. The reason for selecting the electro-hydraulic process must then be that many designs requiring standard type operations have some peculiarity which is not within the capability of conventional equipment, at least, not without some special provisioning or tooling.

The nature of the process is such that large amounts of energy can be directed into isolated areas of the workpiece. One-sixth of an hour is necessary to make the nozzle.

To pierce odd-shaped holes in tubes as shown in Fig. 14.13 would be a rather difficult operation for conventional equipment; at least, if conventional processes were used, the tooling would be intricate and expensive. The operation was quite easily accomplished by electro-hydraulics, using a simple die made from tool steel.

Fig. 14.13. *Die for electro-hydraulic piercing of 1-in. diam tube shown with a stainless and aluminum tube*

When attempting to draw a hemispherical part using press equipment, the center of the blank is contacted first by the punch; pressures then begin to concentrate there and excessive thining ultimately results. Also compression wrinkles in the area of the draw radius are very difficult to eliminate. Electro-hydraulic forming shapes this type of part very easily. To facilitate the forming of this type of part, a wire (to be exploded) can be loaded as a loop in the vicinity of the draw radius. Thus, maximum pressures are created at the draw radius where they are needed, and lower pressure is created at the centre of the blank where equal or greater pressure would cause thining of the blank.

14.6 MAGNETIC-PULSE FORMING

Using this technique, it is possible to apply to metallic workpiece a powerful, uniform impulse solely through the medium of the magnetic field. A pressure of 50000 psi for a period of many microseconds is typical of the magnitude of impulses that are used. Since the magnetic field does not interact appreciably with insulators, the magnetic impulse can be transmitted to the workpiece through the walls of a nonconducting sheath, which may maintain the workpiece in a sterile, an inert, or other special environment. Magnetic impulses have been used to swage and expand tubular forms, as well as to coin, shear, and form flat sheets. To date, the magnetic-pulse process has been primarily applied in the forming of excellent conductors, such as aluminium, copper, brass and low-carbon steel.

General Principles

The basic circuit, consisting of an energy, storage capacitor, a switch, a coil and a power supply that provides energy to charge a capacitor is shown in figure 14.14.

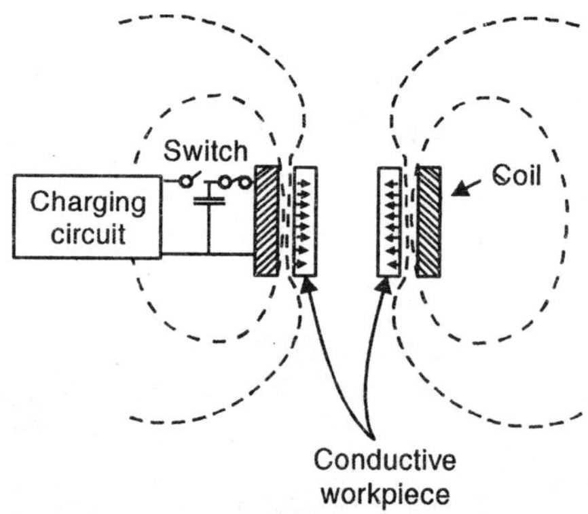

Fig. 14.14. *Basic magnetic pulse metalforming circuit*

In operation, current through the coil produces a magnetic field of high intensity between the coil and the workpiece. During the brief impulse, eddy currents in the work-piece restrict the magnetic field to the surface of the workpiece. The interaction of the magnetic field and the eddy currents creates an inward force on the workpiece as shown in figure 14.14. Assuming that there is zero field on the inner surface of the cylindrical workpiece illustrated, the force on each square centimeter of the workpiece is equal to the energy density in the magnetic field at the surface, that is, equal to $B^2/8\pi$ dynes/cm^2, where B is the flux density in gauss and λ a permeability of unity is assumed. In a forming operation, the magnetic flux density may be as high as 300 kilogauss, which corresponds to a pressure of about 50000 psi. Spiral flat coils and coils of many other configurations may also be used to produce magnetic pulses of various configurations. There are coils used for compression, expansion and forming of flat sheets.

Coil must be designed in such a way that stray inductance is minimised and current concentration are avoided. Since high permissibility materials generally cannot be used in this application because of the high flux density required, field may be ordinarily shaped only through the use of current-carrying conductors. These current-carrying conductors need not be directly connected to the basic coil but may be in inductively coupled configurations similar to the flux-concentrator devices used in induction heating. Figure 14.15 illustrates the use of a massive conducting structure utilizing this technique to concentrate the current, and hence the force, in certain regions of the workpiece thus, relieving the basic coil from the task of carrying high current.

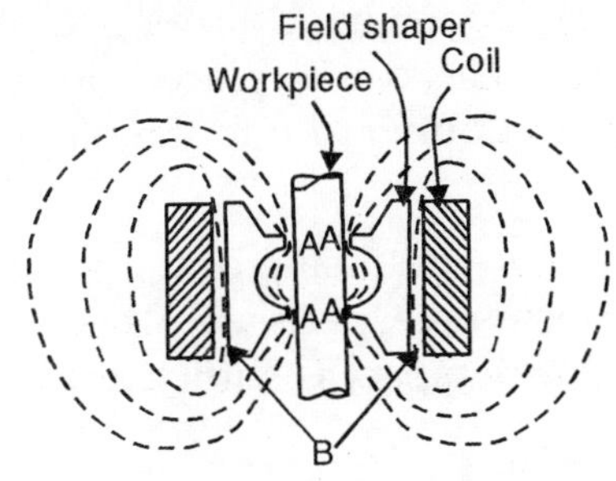

A = high pressure – high current density

B = low pressure – low current density

Fig. 14.15. *Field shaper to concentrate magnetic energy and pressure on two bands of cylindrical workpiece*

Coils and Dies

Coils are classified as being either permanent or expendable. Permanent coils are ruggedly constructed to survive many long production runs. Because of their construction, permanent coils are the most commonly used and also the most expensive to fabricate.

Expendable coils are used for one-of-a kind jobs or for small production runs when the cost of a heavy duty permanent coils is not justified. Expendable coils are designed and constructed to have a lifetime that is comparable with the quantity of parts being produced. This approach reduces coil costs considerably.

Figure 14.16 illustrates the three coil configurations used in electromagnetic forming. The three coil designs and corresponding forming techniques are designated radial compression, radial expansion, and flat coil.

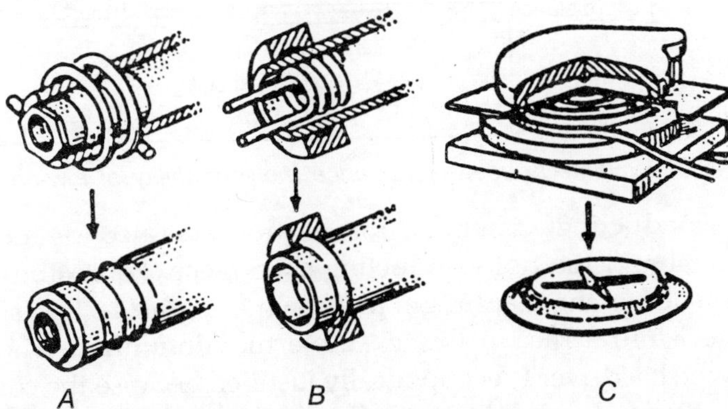

Fig. 14.16. *Electromagnetic forming coils and techniques : (A) Radial core ion forming, (B) Radial expansion forming, (C) Flat coil forming*

Radial compression forming is used both to form cylindrical workpieces and to swage tubes. In practice, radial compression coils are positioned around the outside of the workpiece being formed. The die fits inside. When energy is applied, the workpiece is forced to conform to the die. When used for forming operations, a collapsible die must be used to facilitate removal.

The most commonly used electromagnetic forming technique for cylindrical workpieces is the radial expansion technique. The coil is placed on the inside of the part and when energized causes the part to expand outward into the die. After forming, the die halves are split to remove the part.

The electromagnetic forming or embossing of flat workpieces is achieved by using the flat coil configuration. With this technique, parts are placed on the coil and propelled upward into the die. The velocity of the part impact is sufficient to cause it to conform to the shape of the die cavity.

Whenever possible, dies should be made of nonmetallic materials, but if the strength of metals is required, the die should be made of poor electrical conductors. Otherwise, counter currents that may be generated in the die during the electromagnetic pulse tend to repel the workpiece.

SOME EFFECTS OF RESISTIVITY AND GEOMETRY

The efficiency of magnetic-pulse forming depends upon the resistivity of the metal being formed. It follows that the materials most suitable for magnetic pulse-forming operations are those with high electrical conductivity, such as copper, silver, gold and aluminium which are good, down to mild steel. For good practical results in magnetic-pulse-forming, the resistivity of the material should be less than 15 micro-ohm-centimeters.

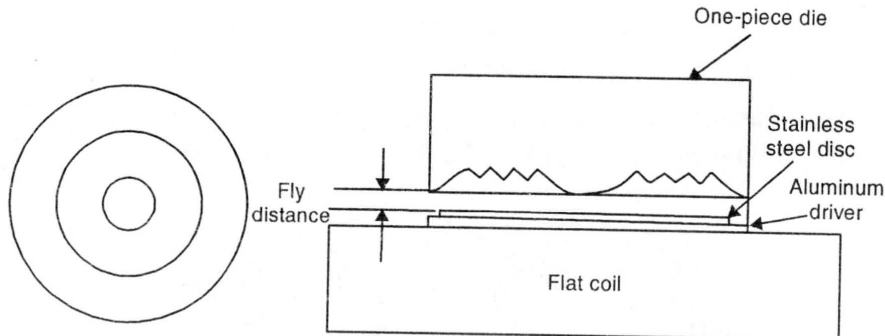

Fig. 14.17. *A consumable driver plate is used to enhance the formability of low-conductivity workpieces*

In working non-conducting materials, forces are generated by using a conductive material between the coil and the non-conducting workpiece. For example, in the forming of a stainless steel part such as the diaphragm as shown in figure 14.17 a sheet of aluminium drives the stainless steel into a die. In this example the aluminium is a driver. There are many applications in which a driver is economically justified because the competing operation is more complex.

In addition to being an electrical conductor, the workpiece must provide a continuous electrical path. The current in a cylindrical workpiece flows around the circumference. Therefore, if a cylindrical workpiece were slit through its length as shown in figure 14.18, the interference with the current flow would reduce and distort the forming forces.

Fig. 14.18. *Cut in cylinder interferes with flow of current and reduces forming forces*

The perforation or cut at end at angle with axis has less effect. Deep slots in the end of tubes, such as shown in figure 14.19 interfere with the current flow in such a way as to produce uneven pressure on the work.

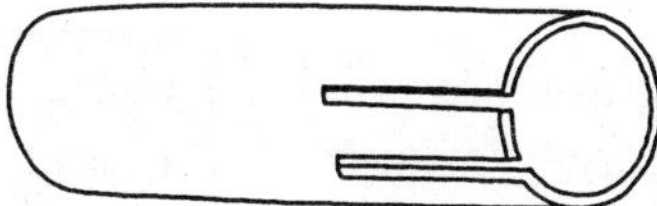

Fig. 14.19. *Deep slots distort action by interfering with flow of current in workpiece*

For efficient operation, the fit of the workpiece to the coil should be considered. Efficiency is best when the gap between the coil or field shaper and the workpiece is a minimum.

APPLICATION EXAMPLES

Electromagnetic forming can be used to form light-gauge metals into shapes up to 300 mm in diameter and in special cases, shape very light gauges in diameters upto 1.3 m. The process is limited to a minimum 25 mm diameter part because of limitations in fabricating small forming coils.

The various applications of Electromagnetic forming are as follows :

Metal to Metal Seal. Metal to metal seals are useful for fabricating pressure vessels, joining lengths of tubing and pipe and assembling such things as coffee pots, tooth paste tubes, turbine shafts and missile nose cones.

Joining Tubes. Figure 14.20 shows a design for a joint between two tubes, using an insert. The insert is made with sharp-edged grooves and should be sufficiently strong to resist permanent deformation.

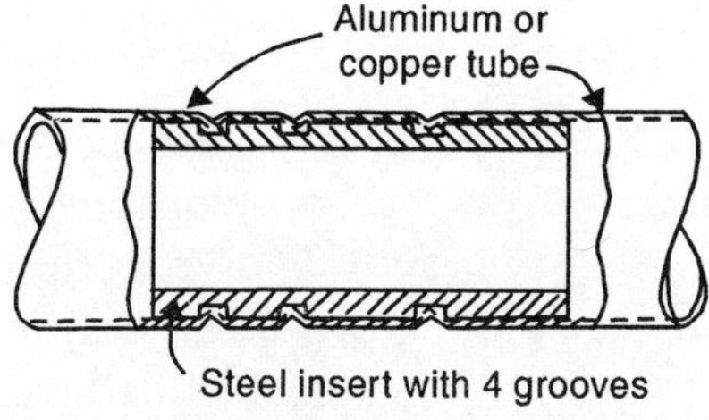

Fig. 14.20. *Typical design of pressure-tight joint between two lengths of tubing*

The joint will benefit if the insert wall thickness is so selected that the insert will elastically deform during the swaging operation.

Banding. Banding includes attaching identification bands, assembling loose parts, applying ornamental devices, and other similar operations.

In applying bands, a relatively thin and highly conductive material is preferred. The band can be initially quite loose on the assembly.

Electrical Connections. Fabrication of electrical connections of superior quality for carrying very heavy currents was one of the early practical jobs accomplished by magnetic-pulse forming. An electrical connection usually requires the swaging of a sleeve over an electrical conductor.

Ceramic Assemblies. Stainless steel, copper, aluminium and brass caps are assembled to ceramic insulators to provide a pressure tight seal by capturing a rubber gasket during the forming of the flange.

Embossing. Ornamental figures are made by using a shallow relief die and a flat coil. In such a figure, there may be a large variance in the degree of forming required for various portions of the figure. For this reason, it may be necessary to use more than one operation.

REVIEW QUESTIONS

1. Distinguish between Electromagnetic Forming and Electro-discharge Forming processes.
2. Give the advantages of high speed forming processes. Mention some typical applications of explosive forming using contact operations and stand-off operations.
3. With the help of a neat sketch explain the principle of working of electro-discharge forming. List the factors that govern the electro-discharge forming of metals and its typical applications.
4. Explain the principle of working of electromagnetic forming with the help of a neat sketch and explain the arrangements for magnetic tube expansion and magnetic pull forming.
5. Describe basic principle, working and general applications of Explosive Forming Process.
6. Write short notes on :

   (*i*) Explosive Compaction.

   (*ii*) Water Hammer Forming

Supplementary Topics

15.0 EXPLOSIVE WELDING

Introduction

The example of explosive welding was observed during World War-I, when it was observed that steel shells of bombs were found to be stuck to metallic objects in the vicinity of the explosion. In 1957 explosive welding was recognized as potential industrial process. Since then, considerable work has been reported in many countries, notably the USA, the USSR, West Germany, Czechoslovakia, Japan and the U.K.

Explosive welding is a solid state welding process wherein coaslescence is effected by high velocity movement produced by a controlled detonation.

Basically, explosive welding involves a high velocity (oblique) impact between (flyer) a plate propelled by an explosive charge and a (second) stationary plate when two plates are to be explosively welded.

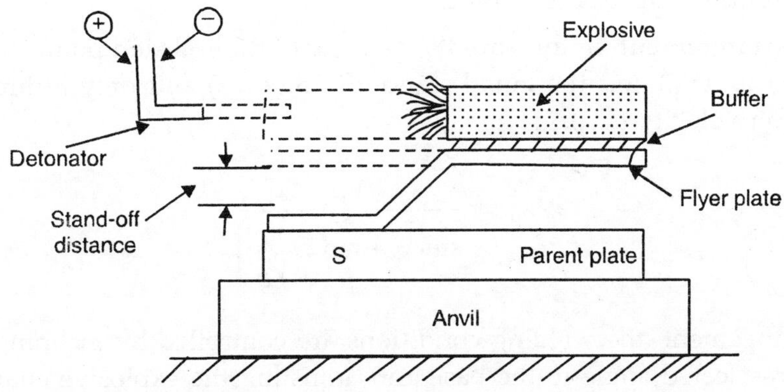

Fig. 15.1. *Parallel arrangement of explosive welding*

MECHANISM OF EXPLOSIVE WELDING (PROCESS PRINCIPLES)

The figure above shows the parallel arrangement of explosive welding. The flyer plate is to be joined with the parent plate. There is a buffer above the flyer plate which may be of rubber, cardboard or similar material to protect the top surface of the flyer plate from damage from detonation of the explosive charge. Above the buffer is a layer of explosive which is detonated from the lower edge. The parent plate rests on an anvil to limit distortion of the final product.

As the explosive is ignited, the detonation wave front progresses across the surface of the flyer plate in a straight-forward and uncomplicated manner. The explosive impulse provides both extremely high normal pressure and a slight, relatively shear or sliding pressure between the flyer plate and the parent plate.

At the point of impact, S, a high instantaneous pressure is generated which is large compared with the shear strength of the materials. Consequently, the metals behave as inviscid fluids which obey the law of fluid mechanics. A thin high velocity jet is formed at S from the surfaces of both plates. This creates fresh virgin surfaces (free from oxides and other films) which are brought together and adhere.

ARRANGEMENT OF EXPLOSIVE WELDING

There are two common arrangements used for producing explosive welds :

(*i*) Parallel (Direct stand-off method)–Also known as contact explosion welding.

(*ii*) Inclined (Angular stand-off method)–Also known as impact explosion welding.

Parallel Arrangement. In this case weld condition are controlled by the stand-off distance, the charge density, the detonation velocity and the deformation characteristics of the flyer plate.

In this arrangement the detonation velocity of the explosive must be less than the velocity of sound in the materials to be welded in order to satisfy the criterion that the collision point velocity must be subsonic.

Inclined Arrangement. In this case the velocity of the collision point (V_{CP}) is a function of the plate velocity (V_P) and the initial stand-off angle (α) and only indirectly dependent on the detonation velocity (V_D).

$$V_{CP} = \frac{V_P}{\sin\left(\alpha + \tan^{-1}\frac{V_P}{V_D}\right)}$$

In this arrangement, the welding conditions are controlled by inclining the flyer plate at a carefully preselected angle to the base and adjusting the explosive charge to drive the plates together at the optimum impact velocity.

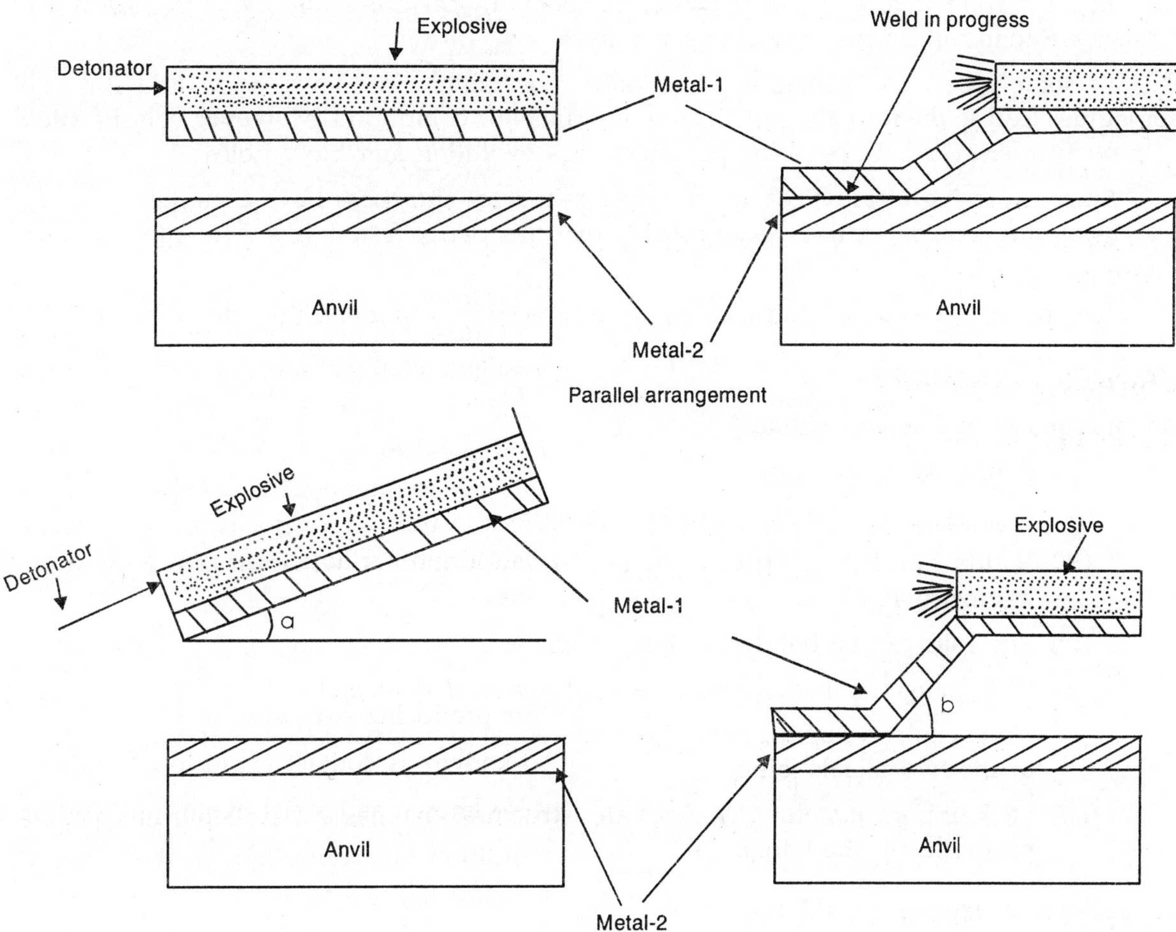

Fig. 15.2. *Inclined arrangement*

Explosive used

Table 15.1

Explosive	Detonation velocity (m/sec)	Density (gm/cm^3)
PETN	8190	1.70
TNT	6600	1.56
RDX	8100	1.65
Tetryl	7800	1.71

METALLURGY OF EXPLOSIVE WELDS

In explosive welding, cleaning of surfaces is not required as it is capable of penetrating great thickness of dirt or oxide. But the resultant accumulation of dirt or oxide near the crests of the ripples may reduce the strength of the joint.

Being a solid-state welding process, the metallurgical problems, therefore, are much less severe than with a fusion welding process.

In some cases, the formation of a continuous molten interlayer occur and this can sometimes lead to the formation of a brittle intermetallic compound. Moreover, solidification of molten interlayer may be accompanied by gas evolution and blow holes.

Since explosive welding results in high strain rates and these have a marked effect in increasing diffusion rates, it is very probable that this partly results in severe alloying with dissimilar metals.

In general, the material after welding are appreciably harder and their ductility reduced.

PROCESS SUMMARY

Advantages of explosive welding

(*i*) Simplicity of the process.

(*ii*) Extremely large surface can be bonded.

(*iii*) Welds can be produced on heat-treated metals without affecting their microstructures.

(*iv*) The foils can be bonded to heavier plates.

(*v*) Wide range of thickness can be explosively clad together.

(*vi*) Good explosive bonds have a strength equal to or greater than that of the weaker of the two metals joined.

(*vii*) Lack of porosity, phase changes and structural changes impart better mechanical properties to the joints.

Limitations of explosive welding

(*i*) In industrial areas the use of explosives will be severely restricted by the noise and ground vibration caused by explosion.

(*ii*) The regulation relating to the storage of explosives may well prove to be the main obstacle to the use of explosive welding.

(*iii*) Metals to be welded by this process must possess some ductility and some impact resistance.

(*iv*) Metal thickness greater than 62 mm of each alloy can not be joined easily and require high explosive loads.

APPLICATIONS OF EXPLOSIVE WELDING

(*i*) Explosive welding has its main applications as that of

(*a*) Welding, (*b*) joining, and (*c*) cladding of metals.

(*ii*) A number of dissimilar metal combinations have been joined successfully with the help of explosive welding. *e.g.* (*a*) aluminium to steel; (*b*) tungsten to steel; (*c*) aluminium to stainless steel etc.

(*iii*) Pipes and tubes upto 1.5 m length have been clad with this process.

(*iv*) It is used die casting industry for nozzles, die-cast biscuits and other components.

(*v*) Heat exchangers tube sheets and pressure vessels (cladding process).

15.1 UNDERWATER WELDING PROCESSES

In the world war, an urgent need was felt for salvaging vessels sunk into the sea and this need raised the status of underwater welding from almost a pipe dream to a practical process.

The international interests to develop and utilize ocean and its resources such as development of offshore gas and oil field, fisheries, large offshore construction etc. have led to the development of underwater welding.

Underwater welding has been used for temporary repair work caused by ship's collisions, unexpected accidents, corrosion and other maintenance works.

PROBLEMS ENCOUNTERED IN UNDERWATER WELDING

(*i*) In underwater welding, the process is same as that of on the surface except that there is :

(*a*) Chilling action of water on the weld metal and the surrounding plate.

(*b*) A higher pressure due to the water-head under which welding takes place.

(*c*) A gaseous envelope surrounding the arc. A gas bubble is formed around the arc due to the combustion of flux on the electrode and the dissociation of water. As welding continues these bubbles, one after the other, travel to the surface of water, thereby making conditions around the arc unstable.

(*d*) An effect on the metallurgical structures and hence on the mechanical properties of the welded joint.

Welding of alloy steels can easily lead to embrittlement in the heat affected zone because of chilling effect of water on the weld metal. Underwater welding is therefore generally restricted to mild steel.

(*ii*) Another problem with underwater welding is that deep water work is never easy, because diving operations are dependent on tide and weather, and the difficulties arising from the various positions in which welding has to be done, also, add to the problem encountered.

(*iii*) An essential requirement for underwater welding is the complete insulation of the welding circuit. Even the electrode coatings are protected by a layer of wax, varnish or cellulose.

Complete insulation is also essential, because the diving suit has many exposed metal parts and the diver in the suit is effectively earthed by the surrounding water. Considering all this, D.C. Welding is employed, as it is safer to use than A.C.

CLASSIFICATION OF UNDERWATER WELDING

(*a*) Wet welding

(*b*) Dry welding

(*i*) Hyperbaric welding

(*ii*) Cavity welding

(*a*) Wet welding

Wet welding is performed by the diver welder normally using surface diving or saturation diving techniques using the MMA (Manual Metal Arc) process with electrodes specially coated with insulating varnishes to keep them dry.

Current is supplied by a generator of some 400 A capacity directly to welding torch head. Interchangeable collets of 4 mm and 4.8 mm hold electrodes of these diameters and are tightened by the twist-grip control, which also used to eject the stub when electrodes are changed. Watertight glands and washers prevents seepage of water into the body of the torch, which is tough rubber covered thus reducing the danger of electric shock.

Arc stability is less than that in air due to the large volumes of gas and steam which are evolved making vision difficult and as a result touch welding is generally used. The presence of water in the immediate vicinity of the weld except at the molten pool under the arc stream results in the very rapid quenching of the weld metal; which produces a hard, narrow heat affected zone and can give rise to severe hydrogen cracking. At present this method is used for non-critical welds, critical welding being carried out under dry conditions.

(*b*) Dry welding

It needs a pressurised enclosure having controlled atmosphere. Weld metal is not in direct contact with water. Dry welding produces very good welds but the process is expensive. Advantages of dry welding are :

(*i*) Reduced hydrogen problem of the environment.

(*ii*) Improvement in stability of welding operation.

(*iii*) Lower weld metal and base metal quench rates.

(*iv*) Weld strength and ductility comparable to surface welds.

(*i*) Hyperbaric welding

Welding is carried out in a dry chamber at pressure above one atmosphere. The chamber is made of steel and are constructed with anti-chambers around the section to be welded. Water is then displaced from the chamber by gas pressure. The gas in the chamber used by the diver-welder is typically a mixture of helium and oxygen. This mixture eliminates the harmful effect of nitrogen, the helium being the carrier for the oxygen–a partial pressure of oxygen above 2 bar can be poisonous. The oxygen content is adjusted until the diver-welder does not use a breathing mask, but at the moment of welding he puts on a breathing-welding mask which has a gas supply separate from that of the chamber and umbilicals connected to the mask to enable the breathing gases to be exhausted outside the chamber.

TIG, MMA and flux cored methods of welding are used down to a depth of 300 m of water but TIG, using high purity argon as the shielding gas, is the slowest method as it is a low deposition rate process. It can be used for root runs and hot passes but MMA or the flux cored method is used for filling and capping. Generally the high strength, basic-coated steel electrodes which may contain iron powder are used. They are pre-heated in a drying oven immediately before use and pre-heat of about 100°C can be applied locally to the joint.

Hyperbaric process has certain limitations :

(*i*) The necessity for seals between the chamber and the structure to be welded is a practical difficulty.

(*ii*) Increase in pressure as depth increases introduces problem both for the welding process and for divers. Weld chemistry, metallurgy and arc physics are all affected.

Air as habitat atmosphere soon becomes objectionable as pressure increases because it presents a potential fire hazard at ambient pressures in excess of 3 atm. Habitat atmospheres of (Helium + Oxygen) and (Argon + Oxygen) are under trials.

(*ii*) Cavity welding

Cavity welding is another approach to weld in a water-free environment. In this process, the conventional arrangements for feeding wire and shielding gas are surrounded by a means of introducing a cavity gas and the whole is surrounded by a trumpet shaped nozzle through which a high velocity water jet passes.

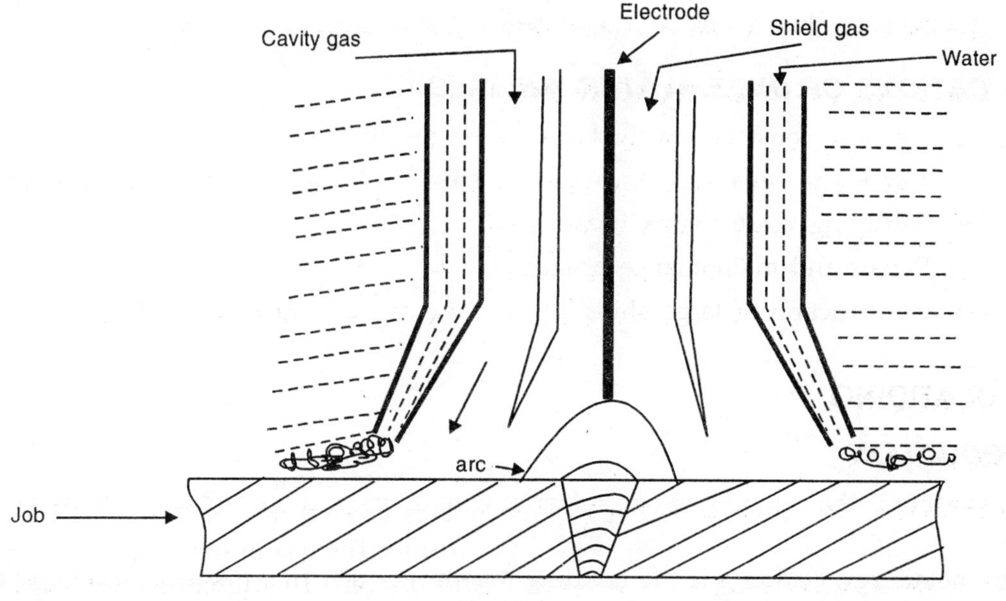

Fig. 15.3

Cavity method avoids the need for a habitat chamber and it lends itself to automatic and remote control.

The process is very suitable for flat structures where butt welds with a back up strip can be welded in the flat or overhead position.

Coffer Dam

A coffer dam is a watertight case of steel piling erected around a given point to keep out the water. By pumping the dam dry, welding can be performed at atmospheric pressure upto about 18 m deep and welding can proceed irrespective of the state of the tides if in tidal water.

In this case the welding is performed exactly as if it were on shore and the welders are in audio and visual contact with the welding process engineer.

CHARACTERISTICS OF A GOOD UNDERWATER WELDING PROCESS

A good underwater welding process should :

(*i*) associate in expensive welding equipment and low welding cost.

(*ii*) decrease electrical hazards.

(*iii*) have 20 cm/min welding speed at least.

(*iv*) permit good visibility.

(*v*) be easy to operate.

(*vi*) produce good quality and reliable welds.

(*vii*) permit welding in all positions.

(*viii*) be such that the operator can support himself by one hand and weld by another hand because of no-gravity in the water.

(*ix*) be such that it can be automated.

APPLICATIONS OF UNDERWATER WELDING

(*i*) Offshore construction for tapping sea resources.

(*ii*) Temporary repair work caused by ship's collisions or unexpected accidents.

(*iii*) Salvaging vessels sunk in the sea.

(*iv*) Repair and maintenance of ships.

(*v*) Construction of large ships beyond the capacity of existing docks.

15.2 CLADDING

INTRODUCTION

Cladding is the covering of one material with another. It has different meanings on the context.

In metallurgy, cladding is the bonding together of dissimilar metals. It is distinct from welding or gluing as a method to fasten the metals together. Cladding is often achieved by extruding two metals through a die or pressing sheets together under high pressure.

The united states mint uses cladding to manufacture coins from different metals. This allows a cheaper metal to be used as a filler.

Regarding optical fibre in telecommunication, cladding is one or more layers of material of lower refractive index, in intimate contact with a core material of higher refractive index.

PROCESS PRINCIPLE

Cladding is usually performed by attaching plates of relatively thick metals together such as by welding and then rolling them down together. The center core and plating metal retain their proportions as the cross-section is reduced.

In hot extrusion process, lubricants should be used between the billet, die and container not only to reduce the work load but also to keep the flow laminar. As a result, the outer surface of the billet forms the skin of the product. This principle of maintaining the surface layer is also a case of cladding.

The temperature range of the billet during the hot extrusion of steels is 1200–1500°C. The die must be kept at a lower temperature (approximately 200°C) to avoid excessive wear rate. Glass fibres (or powder) are normally used as lubricants since the viscosity of glass is sensitive to temp. Thus, the viscosity is high at the die surface, providing a good protection to die wear and facilitating the formation of a glass skin (about 0.025 mm thick) on the product. At the same time, the work load is reduced since the viscosity of glass is much lower at the billet-container interface. This is also a good example of cladding.

Laser cladding is another process for weld surfacing by which a powdered material is melted by use of a laser in order to coat part of a substrate.

The powder used in laser cladding is normally of a metallic nature and is injected into the system by either coaxial or lateral nozzles. The interaction of the metallic powder stream and the laser causes melting to occur and is known as melt pool. This is deposited onto a substrate; moving the substrate allows the melt pool to solidify and thus produces a track of solid metal. This is the common technique, however some processes involve moving the laser/nozzle assembly over a stationary substrate to produce solidified tracks. The motion of the substrate is guided by a CAD system which interpolates solid objects into a set of tracks, thus producing the desired part at the end of the trajectory.

APPLICATION EXAMPLES

Cladding is used :

(*i*) To form new parts.

(*ii*) Apply a harder surface to a part.

(*iii*) Resurface worn or damaged parts.

Since 1943 Metal Cladding Inc. has been proving high-tech coating application for a variety of end uses to enhance functional performance. Whether it be Medical, Industrial, Automotive, Military or other industries, thousands of companies across the USA and Canada rely on the Metal Cladding Inc. to enhance their product performance.

Nuclear reactor fuel

In nuclear reactor, cladding the outer layer of the fuel rods, standing between the coolant and the nuclear fuel. It is made of a corrosion-resistant material with low absorption cross-section for thermal neutrons, usually zircaloy or steel in modern construction, or magnesium oxide with small amount of aluminium and other metals for the now-obsolete magnox reactors. Cladding prevents radioactive fission fragments from escaping the fuel into the coolant and thereby from contaminating it.

Corrosion resistant

Due to superior corrosion resistance of aluminium is added to the strength of duralumin. It is much used in aircraft construction. In a like manner steel may be clad with copper, cupro-nickel, or nickel and rolled gold is made by rolling a gold alloy into a brass or cupro-nickel base. This method in general allows the protection of a material which combines the strength of a core with the corrosion-resistance of the plated material. Unlike electroplating, there is no danger of porosity.

Industrial coatings

Offering high-tech coatings for automotive, chemical processing, food processing, electrical, military, defence. Process include electrocoating, fluoropolymers, powders, high-temperature materials, ultrasonic cleaning, applications of hundreds of spec coatings.

Medical coating

Offering coatings for surgical devices, instruments, disposables, autoclavable hospital and laboratory equipments. Coating applications include conformal coatings, fluoro polymers, antimicrobial materials, liquids, powders, EMI-RFI shielding, speciality cleaning and passivation.

Coil coating

For finned tube heat exchangers, radiators, evaporator/condenser coils and other related HVAC components.

Fuel coatings

Offering coatings for conductive and chemical resistance, hydrophobic, thermally conductive, electrically resistive and diffusion media coating.

Decorative coatings

Offering decorative coatings for metals, composites, plastics glass, ceramics, wood and nearly every solid substrate; coatings include a variety of colour powder coats, sprays and pattern transferred by the sublimation process and water transfer.

REVIEW QUESTIONS

1. Explain the mechanism involved in explosive welding process.
2. Discuss the various types of arrangement made for producing explosive welds.
3. What is cladding ? How it differs from explosive welding ?
4. How cladding operation is performed ?
5. What are application of cladding process ?
6. What are the problems encountered in underwater welding process ?
7. Mention the various types of underwater welding process and its major application.

❑❑❑

Bibliography

ABRASIVE JET MACHINING

1. Sprinborn R.K., *"Nontraditional machining processes"*, ASTME, Michigan, 1967.
2. Sheldon G.L. and Finnie I., *"The mechanics of material removal in the erosive cutting of brittle materials,"* Trans. ASME, Series B, Vol. 88.
3. Lvoie, F.J., *"Abrasive jet-machining"*, Machine design, September 1973.
4. Bhattacharya A., *New technology,* The Institution of Engineers (India), 1973.
5. Sarkar P.K. and Pandey P.C., *"Some investigations on the abrasive jet machining"*, paper presented at the Semi-Annual Meeting, Mech. Engineers Division, Institution of Engineers (India), August 1975.
6. Arion, *"Abrasive machining,* ASTME, Collected papers, Book 6, 1963.
7. Pandey P.C. and Neema M.L., *"Erosion of glass when acted upon by an abrasive jet"*, Proceeding International Conference on Wear, St. Louis, April 1977.
8. Sapra I.L., *"Studies on abrasive jet machining"*, M.E. Dissertation, University of Roorkee, 1975.

ULTRASONIC MACHINING

9. Neppiras E.A., *"Report on ultrasonic machining"*, Metal Work Product; 100(27), 1956.
10. Pentland H. and Extermanis J.A., *"Improving Ultrasonic Machining Rates—Some Feasibility Studies"*, Transactions of ASME, Journal of Engineers for Industry, 1965.
11. Rosenberg L.D., *"Ultrasonic Cutting"*. Consultant Bureau, New York, 1964.
12. Kazamtsev V.F. and Rasenberg L.D., *"The mechanism of ultransonic cutting"*, Ultrasonics, Volume 3, 1965.
13. Markov A.I., *"Ultrasonic Machining of Intractable Materials,* Illife Books Limited, London, 1966.
14. Neppiras E.A., *"Design of Ultrasonic Machine Tools"*, Institutions of Mechanical Engineers, Conference on Technology of Engineering Manufacture, March 1958.
15. Kennedy D.K. and Grieve R.J., *"Ultrasonic Machining—A Review"*, The Production Engineer, Volume 54, September 1975.

WATER JET MACHINING

16. Farmer I.W. and Attewell P.B., *"Rock penetration by high velocity water jets"*, International Journal of Rock Mechanics and Mineral Science, No. 2, Volume 2, 1965.

17. Cooley W.E. and Clipp L.L., *"High pressure water jets for undersea rock excavation"*, Journal of Engineering for Industry, Trans. ASME, Series B, Vol. 92, 1970.

18. Brayon E.L., *"High energy liquid jet as a new concept for wood machining"*, Forest Production Journal, No. 8, Vol. 13, 1963.

19. Neusen K.F. and La Brush E.C., *"Material removal by high pressure liquid jets at ten kilobars"*, Journal of Engineering for Industry. Trans. ASME, Series B, Vol. 97, 1975.

ELECTRO-CHEMICAL MACHINING

20. Allison C.R., *"How Electrolytes Influence the ECM Process"*. Creative manufacturing Seminars, American Society of Tool and Manufacturing Engineers, Detroit, Michigan, U.S.A., 1963-64.

21. Benton R.C. and Woodring G.D., *"chemical machining"*, American Society of Tool and Manufacturing Engineers, U.S.A.

22. Sorkhel S. Mishra, P.K. and Sur. B., *"Determination of Optimum Working Parameters in Electro-chemical Machining"*, Technical Session on Electro-chemical Instrumentation of the society for Advancement of Electro-chemical Science and Technology, 1972.

23. Sorkhel S. and Sur, B., *"Some Aspects of Electro-chemical Machining"*, Journal of the Institution of Engineers (India), Vol. 53, 1972.

24. Sorkhel S. Mishra, P.K. and Bhattacharya, A., *"Electro-chemical Machining"*, Proceeding of Fifth All India MTDR Conference, Roorkee, 1972.

25. Sorkhel S. and Sur, B., *"Mechanism of Electro-chemical Grinding"*, Journal of the Institution of Engineers (India), Vol. 53, 1972.

26. Tipton H., *"The Dynamics of Electrochemical Machining"*, Advances in Machine Tool Design and Research, "Oxford Pergamon Press, Ltd.

27. Wilkinson B., and P. Warburton, *"Electro-chemical Machining"*, Proce. Conference on Machinability, Iron and Steel Institute, London, 1965.

28. Maleka A.H., *"The application of electrochemical machining"*, Proc. Conference on Machinability, Iron and Steel Institute, London, 1965.

29. Cole R.R., *"Basic research in electrochemical machining—Present status and future directions"*, International Journal of Production Research, Vol. 4, 1965.

30. Brandi R.R., *"Basics of Electrochemical Grinding"*, American Machinist, No. 9, Vol. 118, 1974.

31. Koeing W. and Degenhardt H., *"Electrochemical machining of heat resistant alloys"*, Annals, CIRP, No. 1, Vol. 21, 1972.

32. Landolt *"Mechanical aspects of electrochemical machining of metals"* (in German), Chem. Ing. Tech., No. 4, Vol. 45, 1973.

33. Loutrel S.P. and Cook N.H., *"High rate electrochemical machining"*, ASME Paper, May, 1973.

ELECTRO-DISCHARGE MACHINING

34. American Machinist's Special Report, "*Cutting with a Spark*". Report No. 590, July 1966.

35. Banerjee, P.P., Mishra, P.K. and Bhattacharya, A. "*Accuracy of Machining by EDM Process*". 17th Congress of Theoretical and Applied Mechanics Congress, Ranchi, 1972.

36. A.N. Chakrabarty and Bhattacharya, A., "*Effect of Dielectric Medium on the Performance of EDM Process*". Journal of the Institution of Engineers (India), Vol. 51, 1970.

37. Chaudhary, S. Das, A.K. and Bhattacharya, A., "*Determination of Critical Resistance in Spark Erosion Machining for various Electrodes in a given Dielectric*". Proceedings of Second All India MTDR Conference, India, 1968.

38. Cook, N.H., "*Manufacturing Analysis*", Addison Wesley.

39. Davis, M.F., "*A Synopsis of Comprehensive Investigation in EDM*". Creative Manufacturing Seminars, American Society of Tool and Manufacturing Engineers.

40. Kurafuzi, H. and Suda, K., "*Study on Electrical Discharge Machining*". Proceeding of First All India MTDR Conference, Jadavpur, 1967.

41. Lazarenko, B.R., "*The Present State of Development of Electro-spark Machining of Conductive Materials Abroad*". Electro-spark Machining of Metals, Vol. 2, Edited by B.R. Lazarenko.

42. Pal, M.N., Mishra P.K. and Bhattacharya, A., "*Optimization of Circuit Parameters of Relaxation Circuit of EDM*". International Conference on Non-traditional Processes, Timisoara, Rumania, 1971.

43. Barash, M.M. "*Electric Machining of Metals*", Modern Workshop Technology, Part-II, Editor H. Wright Baker, Second Edition, 1960.

44. Blake, L.R., "*High power spark erosion machine*", The Engineer, No. 5159, Vol. 199, 1955.

45. Dave, B.J., *Servo Controlled Electrical Discharge Machines*, Proc. IV, AIMTDR Conference, 1970.

PLASMA ARC CUTTING

46. "Plasma Arc Machining", Report No. AMRA-CR 66-03 F, US. Army Report.

47. Hebble C.M. Jr, "*Plasma cutting—an exotic process comes into its own*", Welding Engineering, No. 3, Vol. 58, 1973.

48. Nachman M., "*Few considerations about the designing of plasma are torches*", Rev. Int. Hautes Temp., Refract, No. 2, Vol. 10, 1973.

49. Kharitonov E.P., "Magnetic control of plasma are during the cutting of metals", Weld Prod. No. 12, Vol. 19, 1972.

50. Budnik N.M., "Plasma spraying of a protective coating from aluminia on refractory materials", Weld Prod., No. 12, Vol. 20, 1973.

51. Ingham H.S., Jr and Fabel A.J., *"Comparison of plasma flame spray gases"*, Weld Journal, No. 2, Vol. 54, 1975.

52. Tbilisi Division of VNIIESO, *"plasma welding equipment"*, Weld Prod. No. 11, Vol. 20, 1973.

53. Beider B.D. *et al.*, *"Determination of the parameters of a plasma arc in a stream of argon and nitrogen used for the cutting of metals"*, Weld Prod. No. 6; Vol. 21, 1974.

54. Larri O. Ya, *"Equipment for the air plasma cutting of metals"*, Weld Prod. No. 5, Vol. 21, 1974.

55. Shapiro I.S. *et al.*, *"The influence of parameters of a gas supply system on the operating characteristics of plasmatrons for metal cutting"*, Weld. Prod., No. 7, Vol. 23, 1976.

56. Rao K.N. and Gururaja G.I., *"Some Studies of the Plasma Cutting of Aluminium Plates"*, Indian Welding Journal, No. 3, Vol. 6, 1974.

ELECTRON BEAM MACHINING

57. Christy R.W., *"Formation of thin polymer films by electron bombardment"*, Journal of Applied Physics, Vol. 31, 1960.

58. Yoshida S., *"Radioactive isotope study of the dissociation of barium wide under electron bombardment"*, Journal of Applied Physics, Vol. 27, 1956.

59. Baker A.G. and Morris W.C., *"Decomposition of metallic films by electron impact decomposition of organometallic vapours"*, Review Scientific Instruments, Volume 32, 1961.

60. Anon, *"Some elementary considerations in design of electron beams for welding and heating"*, Proc. 3rd Symposium on Processes, Boston, 1961.

61. Anon, *"Electron beam processing"*, Aircraft Production, No. 8, Volume 22, 1960.

62. Lamba S., "Applications of Electron Beams for Evaporations and Machining of Materials,'Proc. 4th Symposium on Electron Beam Technology, Boston, 1962.

63. Duhamel R.F., Practical Electron Beam Cutting and Milling Application, ASTME Creative Manufacturing Seminars, 1962.

64. Kuper G.I., *"Electron beam device for removal of dielectric and metallic materials"*. VDIZ, No. 5, Vol. 116, 1974.

65. Miyazaki T. *"characteristics of electron beam drilling of metals"*, Bulletin Japan Society of Precision Engg. No. 2, Vol. 10, 1976.

LASER BEAM PROCESSES

66. Bhattacharya A., *"Laser Features and Application"*, Journal of Metal Progress, Jan. 1966.

67. Bhattacharya A., *"Laser for Machining Applications"*, Journal of the Institution of Engineers (India), Vol. 4, 1971.

68. Bhattacharya A., and Mullick B.K., "*Machining and Welding by Laser Beam*", Journal of the Institutions of Engineers (India), Vol. 51, 1971.

69. Fairbanks R.H and Adams C.H., "*Laser Beam Fusion Welding*", Welding Journal, Vol. 43, 1964.

70. Gagliano F.P., Lumley R.M. and Watkins L.S., "*Lasers in Industry*", Proceedings of the IEEE, Vol. 57, 1969.

71. Lawrence T.R., "*Lasers*", Journal of Washington Academy of Sciences, Vol. 53, 1963.

72. Maiman T.H., "*Stimulated Optical Radiation in Ruby*", Nature, Vol. 187, 1960.

73. Franken Peter, "*High-Energy Lasers*", International Science and Technology, 1962.

74. Schawlon L, "*Optical Masers*", Scientific American, 1961.

75. Schmidt A.D., Ham I. and Hosi T., "An Evaluation of Laser Performance in Microwelding; Welding Journal, Vol. 44, 1965.

76. Williams D.A., "*The Laser as a Drilling Tool*", Pennsylvania State University Proceedings, 1966.

77. Young D.S., "*The Laser as an Industrial Machining and Welding Tool*"; Pennsylvania State University Proceedings, 1966.

HIGH ENERGY RATE FORMING

78. Strohecker D.E. and Dwens D.H., "*A Guide to the Literature on High Velocity Forming*", Report No. 179, Defence Metals Information Center, Battelle Memorial Institute, 1962.

79. Suits C.G., "*Notes on High-Intensity Sound Waves*", General Electric Review, 1936.

80. Kapitsa P, "*A Method of Producing Strong Magnetic Fields*", Proc. Roy. Soc. A., 1924.

81. Langlois A.P., "*Metal Forming with Electromagnetics*", ASTME Creative Manufacturing Seminars, 1960.

82. Clark D.S., and Wood D.S., "*The Time Delay for the Initiation of Plastic Deformation at Rapidly Applied Constant Stress*", Proc ASTM, 1949.

83. Lingon R.E. and Cruver R.W., "*Explosive Forming with Gas Mixtures*", ASTME Creative Manufacturing Seminars, 1962.

84. Wagner H.J., and Sabroff A.M., "*The History and Scope of High Energy Rate Forming*", ASTME Creative Manufacturing Seminars, 1960.

85. Zernow L., "*Application of High Velocity Metal Forming in Short Run Production*", ASTME Creative Manufacturing Seminars, 1962.

86. Wood W.W., "*Applications and Limitations of High Energy Rate Forming Processes*", Technical Report 2-22300/2R-18, Chance Vought Corporation, Dallas, Texas Aug. 1962.

87. Dieter G.E., "*Strain-Rate Effects in Deformation Processing*", Ninth Sagamore Ordnance Materials Research Conference, Raquette Lake, New York, 1962.

88. *"Sheet Metal Forming Technology"*, ASD Interim Report 7-871 (III), Contract AF 33, Chance Vought Corporation, Dallas, Texas, 1962.
89. Broberg K.B., *"Shock Waves in Elastic and Elastic-plastic Media"*, Private Printing, Stockholm, 1956.
90. Pearson J. *"Metal Cutting with Explosive"*, ASTME Creative Manufacturing Seminars, 1961.
91. Pearson J. and Hayes G.A., *"Some Material Behaviour Patterns in Explosive Working"*, ASTME Creative Manufacturing Seminars, 1962.
92. Cole R.H. *"Underwater Explosions"*, Princeton University Press, Princeton, N.J., 1948.
93. Underwater Explosion Research, Vol. 1, The Pressure Pube (a compilation of paper), Office of Naval Research, 1948.
94. Rinehart J.S. and J. Pearson, Explosive Working of Metals, Pergamon Press, London, 1962.
95. Roth J., *"The Forming of Metals by Explosive"*, The Explosives Engineers, 1959.
96. Rinehart J.S. and Pearson J., *"Behaviour of Metals under Impulsive Loads"*, American Society of Metals, 1954.

Index

F

G

I

J

L

M

□□□